PREPARING FOR GENERAL PHYSICS

Math Skill Drills and Other Useful Help

Arnold D. Pickar
Portland State University

Addison-Wesley Publishing Company
Reading, Massachusetts • Menlo Park, California • New York
Don Mills, Ontario • Wokingham, England • Amsterdam • Bonn • Sydney
Singapore • Tokyo • Madrid • San Juan • Milan • Paris

Library of Congress Cataloging-in-Publication Data

Pickar, Arnold D.
    Preparing for general physics : math skill drills, and other
useful help / Arnold D. Pickar.
            p. cm.
        Includes index.
        ISBN 0-201-56952-3 : $12.95
        1. Mathematics.   I. Title.
QA39.2.P56   1991
510--dc20   91-40685
                                                               CIP

Reprinted with corrections September, 1993

ISBN 0-201-56952-3
    4 5 6 7 8 9 10-AL-95949392

Table of Contents

To The Instructor

Although this book is primarily meant to be used for self study, it is an outgrowth of my experience teaching abbreviated courses in remedial and preparatory skills for students about to enroll in general physics classes. In teaching these classes, my greatest frustration arose out of a lack of suitable text materials.

Whereas several books have been published in the past aimed at such students, and some of them have much to commend them, none quite serve all the important objectives of skill preparation as I see them. This is especially critical if the book is to serve students trying to prepare without benefit of a formal class. Some of the books are too verbose, or too inclusive, or too mathematically oriented. Above all, the lack of sufficient and appropriate solved practice problems is a common difficulty. A great deal of the material taught in prerequisite math courses is not needed in beginning physics courses. What *is* needed is guidance and lots of drill in a context which emphasizes the use, rather than the theory, of the mathematics. Just as important, a suitable text, by exercising students in appropriate verbal and pictorial skills, should help develop in them the art of thinking about physics.

The primary audience for this book is meant to be students about to enter or just beginning a college non-calculus level course in general physics, especially those students whose basic mathematical skills are weak or dated . However all of the same skills required by these students are needed, and are often poorly developed, in students attempting calculus level physics; for them, working with this book would certainly not be detrimental.

Arnold D. Pickar
Portland, Oregon

To the Student

IS THIS BOOK FOR YOU?

If you are about to begin a college course in introductory physics for science majors, or perhaps have already begun, you may be feeling anxious about what you are getting into. If that is the case, this book may be for you.

For many college students physics appears to be a "different" type of course. Unlike so much of the learning you have had to master over the years — even in science courses — the material in a college-level physics course minimizes the importance of memorizing of facts and standard recipes for solving problems. The emphasis is on understanding concepts and the ability to apply this understanding in solving a wide range of problems.

The obstacles to mastering the art of solving physics problems are many but you can overcome them. Factors which lead to success include the following:

- the competence to use mathematics (usually fairly simple mathematics) with ease and accuracy;

- the ability to visualize the situation a problem describes and the ability to comprehend what is being asked;

- the knowledge required to translate, when necessary, the words of a problem and images they bring to mind into an appropriate mathematical form;

- the sensitivity required to distinguish between those problems which do not require a mathematical solution from those which do;

- the confidence that the use of unfamiliar words will not be an obstacle to learning new ideas.

This is a list of basic skills and habits of thought which hopefully you have developed in the past through contact with other science and mathematics courses. But unless you are completely confident that you have achieved these skills and habits, going through this book is likely to prove worthwhile.

You should recognize that this is *not* a book whose primary purpose is to teach you physics or to introduce you for the first time to the prerequisite skills. Rather, it is a book to help you hone the skills you already have in order to make the study of physics as effective a possible. The experience of physics is both beautiful and challenging; the challenge is most successfully met, and the beauty most enjoyed when the student is not distracted and inhibited by a less than adequate mastery of basic skills.

IF THIS BOOK IS FOR ME, WHEN SHOULD I BEGIN — AND HOW LONG WILL IT TAKE?

The best time to begin preparing for general physics using this book is 4 to 6 weeks before the course begins. However if you should realize only later that you need this preparation — even after the course has begun — you cannot help but profit from working with this book, providing you are willing to devote the time and effort necessary to do most of it before the physics course gets too far along.*

You can probably finish this book with 20 to 30 hours of work. Preferably this should be done in sessions of several hours, spread out evenly over as much as a month or, if necessary, as little as a week. Moreover, the several pretests may indicate you are already strong in some aspects of the work and can go quickly over the related material. On the other hand, if the post-test (at the end of each "Round" of work) indicates a continuing weakness with some skills, you may wish to review some of the material again.

But remember — whatever the amount of time you require to get through this preparation for physics, don't put it off. Physics courses start out in full stride — and every step forward depends to a greater or lesser extent upon the ones that precede it. The object of "Preparing for General Physics" is to assist you in keeping up with the course so that you never get behind.

HOW THIS BOOK IS ORGANIZED AND HOW TO WORK WITH IT

1. Take the pretest.

Each of the five major sections ("Rounds") in this book begins with a timed pretest. Taking this test should give you an understanding of your strengths and weaknesses with respect to the skills emphasized in that round of material. In the reviews and drills that follow you can pay special attention to those areas in which you are weakest. However it would not be unwise to do most of the drills, including ones related to skills you seem to have well in hand. This can only improve your speed and accuracy and help you apply the material in a physics context.

2. Review.

Each round of material is broken into several skill areas for which a brief Review, including fully explained examples, is given. You will discover that many of the examples have a physics-like context. But not to worry — the emphasis is on the basic skills, not on any physics to which the problem may be related. The purpose is to reinforce your ability to think about word problems, while familiarizing you with some of the vocabulary of physics.

(There is also a short essay on physics following each Round which you may find interesting to read. These are not required for skills preparation, but they can help give you some insight into the subject in which you are about to invest so much of your time and effort.)

3. Do the "skill drill".

This is the heart of your preparation for general physics. Space is provided beneath or alongside each problem for your solution. After you have done your best to answer some of the problems on one page, you may compare your own responses with the book's solutions; simply turn over the page and look at the reverse side. Avoid the temptation to look at a solution before you have tried to find your own solution.

---

*NOTE: The last part of the book (Round Five) can be delayed, if necessary, since the material it contains is not usually needed during the first two months of a typical physics course. However it is wise to finish this work long before you actually need it.

Examine the suggested solutions carefully. Do you understand the methods used? How do your answers compare with the book's answers? Are you ready to go on to more drill questions, or should you look back at the Review section? If you are satisfied with your understanding of the problems you have just completed, go on to the next questions, trying to build up speed without sacrificing accuracy. Continue until you have finished all the drill problems for that skill area.

4. Take the post-test.

After you have finished the drills for all the skills areas in a particular Round of work, take the post-test. Try to do this in the allotted amount of time; then check the answers at the end of the test.

The results of the post-test should show whether you need to enhance some skill areas. Go back and review the pertinent discussions; try to do the example problems on your own; work certain drill and test problems again. Then, when you are ready, move on to the next round of skill areas until you have completed all five rounds.

TOOLS OF THE TRADE

Before you start working, supply yourself with the following:

1. **A calculator.** Much of the numerical work in the Drills (but not all of it) can be aided using a handheld electronic calculator. An *inexpensive scientific calculator* which can handle numbers in both decimal and exponential form is recommended. Besides the usual arithmetic operations, you will need at least $\sqrt{x}$, **sinx**, **cosx**, **tanx**, **sin^{-1}x**, **cos^{-1}x**, and **tan^{-1}x** functions. Also **1/x** and **x^2** are useful. For the last Round **e^x**, **logx**, and **lnx** keys are required.

2. **Scratch paper and pencil**. Although space is provided for your work in the drills, it is helpful (as it will be with actual physics assignments) to use scratch paper to help yourself organize your thoughts and make preliminary attempts. (In an actual physics course, the solutions you submit for grading should be neat and organized — this book's solutions are meant to be a reasonable model.) Except perhaps for laboratory work, pencil is the writing instrument of choice. Neat erasures are preferable to ugly scratching-out, so be sure to have a good eraser.

FINAL WORDS BEFORE YOU BEGIN

In doing the work outlined above you should keep in mind what are the goals and what are not the goals of this book:

- This book is not meant to be a physics text. Moreover, although ideas and concepts of physics are referred to in many places, you need not to have studied physics previously.

- The use of this book is not meant to substitute for the mathematics courses which are prerequisites for your physics course. The mathematics reviewed here is much more restricted in scope and is treated with far less rigor. The emphasis is on enhancing your *facility* with mathematics skills frequently used in a beginning physics course.

- Although the organization of this book is largely built around a sequence of mathematical topics, the skills which are useful in doing physics encompass more than pure mathematics. Among other things the reviews and drills in this book should help you deal with word problems, think pictorially, make diagrams, use short cuts, and be discriminating about your methods and answers.

Physics is a fascinating science which is fundamental to all other sciences — and you can master it if you work at it seriously and do not get behind. You are more likely to keep up your progress if you are not distracted and slowed down by difficulties which have little to do with the physics per se.

> *To help you build up the background which will permit you to concentrate on the physics itself is the purpose of this book.*

Now go to the next page and begin!

Round I — Words and Numbers

The first part of this book deals with very basic skills which you will need over and over throughout your study of physics. Some of these things—doing arithmetic, interpreting word statements, using units—are probably very familiar to you, as they are taught from the earliest grades in school. Some related topics may not be as familiar. They may even be new to you. The pretest below will give you an opportunity to see whether you can deal with this sort of material with comfort, speed, and accuracy.

PRETEST — Optimum test time: 18 minutes or less.

Have paper, pencil, and calculator on hand, although these are not required for every question. Note your starting time, then work each problem, putting your answer in the space provided. When you are done, note your ending time before checking your results against the answers given following the test.

STARTING TIME_____   ANSWERS

1. Express 15,621 using scientific notation assuming the number is accurate to three significant figures.   _____

2. Express 2.34×10^{-3} in decimal form.   _____

3. Do an "order of magnitude" mental calculation (no paper, pencil, or calculator) to find the range of values in which the following expression lies: _____

$\dfrac{(7.15)(8604)}{0.0702}$ lies between
    (a) 1,000 and 10,000
    (b) 10,000 and 100,000
    (c) 100,000 and 1,000,000
    (d) 1,000,000 and 10,000,000

4. Without using a calculator quickly find the approximate value (within a factor of two) of the following expression:

$\dfrac{(4.52)(76.1)}{(581)(0.413)}$ _____

5. Use a calculator to evaluate the following numerical expression. Your answer should be expressed in scientific notation to the appropriate number of significant figures:

$\dfrac{(160.)(3.169 \times 10^{3})}{72.}$ _____

6. Without a calculator determine the value of $\sqrt{90. \times 10^{-5}}$.   _____

7. In a European supermarket you buy 10 "kilos" (kilograms) of sugar. You might guess the weight of the sugar to be close to:

(a) 10 pounds (b) 20 pounds (c) 50 pounds (d) 100 pounds

_____

8. You drive for 2 hours at a steady speed of 55 miles per hour. If you use up 5 gallons of gasoline, what is your fuel mileage (in miles per gallon)?

_____

9. How many feet are there in 3.0 meters? (1 m = 39.37 in, 12 in = 1 ft.) You may use a calculator.

_____

10. Write each of the following statements as an algebraic equation:

(a) Bob is 8 years older than Joe was 3 years ago. (Use B and J to represent the present ages of Bob and Joe in years.)

_____

(b) If Ed were 20 pounds heavier he would be twice as heavy as Alice.

_____

11. Make a simple sketch which represents the following situation: a chair tipped back on its rear legs so that they make an angle θ with respect to the floor.

12. Sketch a freehand map which illustrates the following statement. Label the drawing and show the line of sight from girl to mountain.

Having started out from her camp (C), Joan (J) walked in a north-easterly direction until after a mile she spotted the summit of Old Baldy (B) lying about 2 miles directly south.

13. A platoon of six soldiers line up in order of height. Dan, Joan, and Carl are taller than Emily, Sam, and Henry. Joan is taller than Dan, but shorter than Carl, whereas Sam is taller than Henry who, in turn, is taller than Emily. In what order do they line up? (HINT: A diagram may be useful for finding the answer.)

ENDING TIME_____

_____

ANSWERS:

1. 1.56×10^4

2. 0.00234

3. (c)— around 90,000

4. 1.5

5. 7.0×10^3

6. 3.0×10^{-2}

7. (b)— about 22 pounds

8. 22 mpg

9. 9.8 ft

10. B = (J-3 yr) + 8yr

    E + 20 lb = 2A

11.

12.

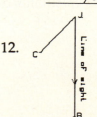

13. CJDSHE

Review 1 — Dealing with Numbers

If there is something which characterizes physics when compared with other fields of study, including its sister sciences, it is an extraordinary reliance on numbers to describe and verify conclusions. In this section we review important ways of writing numbers and how numbers can be efficiently combined to give answers with an appropriate degree of precision.

SCIENTIFIC NOTATION

Physics involves concepts which are described by numbers ranging from the unimaginably small to the astronomically large. The most convenient way to express numbers over a wide range is called "scientific notation."

In scientific notation numbers are represented by the product of a multiplying factor and a power of ten.

(A "power of ten" is the number 10 raised to an integer exponent.) Negative as well as positive multiplying factors and exponents may be used. Some examples of numbers written in scientific notation and their ordinary decimal equivalents are the following:

$$35.00 \times 10^6 = 35{,}000{,}000$$
$$-2.70 \times 10^3 = -2700$$
$$4.3 \times 10^{-2} = 0.043$$

Besides being a compact way of writing quantities over a wide range of values, scientific notation also gives some idea of how precisely a quantity is known. All the digits in the multiplying factor in front of the power of ten are considered *significant* — they convey meaningful information and are to be "taken seriously." More will be said about "significant figures" later on.

Translating into scientific notation and back again. The above examples illustrate the meaning of the power of ten: 10^6 is simply a way of writing 1,000,000, 10^3 means 1000, and 10^{-2} means 0.01.

Given a number in decimal form, to find the power of ten in its scientific notation equivalent: the exponent is the number of places from where you wish to place the decimal point in the multiplying factor to where it lies in the decimal form.

The sign of the exponent depends on whether you count off places to the right (+) or left (−). How this procedure is used to work out one of the numerical examples above is discussed here:

Write 2700 in scientific notation, assuming 3 significant figures.

DISCUSSION: The multiplying factor is written with 3 digits to indicate three significant figures; a convenient value is 2.70 . (We could have also chosen 0.270, or 27.0, etc.; the power of ten simply would have to be different.) Counting off places to the *right* this way:

$$2 \overset{1}{\frown} 7 \overset{2}{\frown} 0 \overset{3}{\frown} 0. \rightarrow \text{ the exponent in the power of ten is +3.}$$

Thus the scientific notation equivalent is 2.70×10^3.

Here is a second example in which the scientific notation equivalent contains a *negative* exponent in the power of ten:

Write 0.043 in scientific notation, assuming 2 significant figures.

DISCUSSION: Choose the multiplying factor to be 4.3. Then counting off to the *left* this way:

$$0 . \overset{2}{0} \overset{1}{4} 3 \rightarrow \text{ the exponent is } -2.$$

The equivalent is thus 4.3×10^{-2}.

To go from scientific notation to a decimal form the above procedure is reversed, as in this example:

Write 35.00×10^6 in decimal form.

DISCUSSION: Move the decimal point in the multiplying factor 6 places to the *right*, as follows.(In other words multiply by 1,000,000.)

$$3 5 . \overset{1}{0} \overset{2}{0} \overset{3}{0} \overset{4}{0} \overset{5}{0} \overset{6}{0} \rightarrow 35,000,000.$$

The final result is ambiguous about the number of significant figures.

CALCULATIONS AND ESTIMATES

Calculators. The use of electronic calculators is encouraged for finding numerical answers to physics problems, but this should be done only when appropriate and only with intelligence. Operating a calculator is not entirely error-free, but mistakes can often be caught if one has the habit of making quick estimates.

When your answer looks ridiculous it is wise to recheck your logic and your calculation!

Approximate answers. Scientific notation makes it simple to multiply and divide numbers to get an approximate answer without a calculator. As an example consider the following combination of three numbers:

$$\frac{(2700)(0.043)}{35,000,000} = ?$$

To get a quick estimate, the expression can be rewritten in scientific notation with each multiplying factor rounded off to a single digit. The calculation is then carried out using the usual rules for multiplying and dividing, as follows:

The multiplying factors are separately combined to find the multiplying factor in the answer. To find the exponent in the answer the individual exponents are added if there is a multiplication or subtracted if there is a division.

Thus for the combination of numbers given above

$$\frac{(2.70\times10^3)(4.3\times10^{-2})}{3.500\times10^7} \approx \frac{(3)(\cancel{4})}{\cancel{4}}\times10^{(3-2-7)} = 3\times10^{-6}$$

(The symbol $\approx$ means "approximately equal." The "slash marks" / indicate that the 4's divide exactly into one another to give a value of 1, i.e., they "cancel.")

Order of magnitude estimates. With practice, getting a sense of the size of an answer by doing rough calculations can become routine and often can be done in one's head. This is particularly true when only an "order of magnitude" answer is wanted. An order of magnitude estimate is usually considered one in which the exact value is rounded off to the nearest factor of ten. Thus an order of magnitude result for the above calculation is

$$\frac{(2700)(0.043)}{35,000,000} \sim (10^3)(10^{-2})(10^{-7}) = 10^{-6}$$

(The symbol $\sim$ stands for "is order of magnitude of.")

SIGNIFICANT FIGURES

The number of digits used to write out a number can tell us how precise the number is meant to be. For instance something described by the number 2.54 can be supposed to have an actual value between 2.535 and 2.545. But simply writing 2.54 means we are uncertain about just where in that range the actual value falls. Only three figures are "significant."

Calculations and significant figures. When 2.54 is multiplied or divided by a number which is more precisely known, the result is only meaningful to three significant figures. For example, the most accurate value we can give for the product (2.54)(3.213) is 8.16. It is not 8.16102, the number which a calculator might display.

> *When multiplying or dividing, the answer cannot be more precise than the least precise <u>factor</u> in the calculation; this usually means that the answer has the same number of digits as the factor with the least number of significant figures.*

(For numbers somewhat greater than a multiple of ten, an extra digit does not imply greater precision. For instance, the three digit number 105, is just about as precise as the two digit number 95.)

In any event, do not confuse the number of decimal places with the number of significant figures. The number 0.0254 has the same number of significant figures as does 2.54; both should be regarded as being known to about 1 part in 254.

> *It is bad form as well as incorrect to write down all the digits displayed by your calculator. Your answer should be rounded off to the appropriate number of significant figures.*

Exact numbers. There is an exception to the above rule about significant figures of products and quotients. Some numbers have, by implication, an unlimited accuracy, and therefore cannot set a limit on the number of significant figures in the answer. For example, suppose we wish to calculate the diameter of a circle, given its radius. If the radius is 3.4 inches then the correct calculation is:

diameter = (2)(3.4 inches) = 6.8 inches.

Although the factor of 2 is written with only one digit, it might be thought of as having an unlimited number of significant figures (2.00000...). Another example is the irrational number $\pi = 3.14159...$; in any calculation, π can be written out, in principle, to as high degree of accuracy as is needed.

CALCULATING POWERS AND ROOTS

Squares and higher powers. These often appear in physics calculations. If a number is expressed in scientific notation it can be raised to a power n using the following rule:

> *First the multiplying factor is raised to the power n to give the multiplying factor of the answer; then the exponent in the power of ten is multiplied by n to give the exponent in the answer.*

That this rule follows from the rule for multiplying numbers expressed in scientific notation is illustrated by this example:

Take the square of 3×10^3, i.e., find $(3 \times 10^3)^2$.

DISCUSSION: The square can be written

$$(3 \times 10^3)(3 \times 10^3) = (3)(3) \times (10^3)(10^3) = 9 \times 10^{(3+3)}$$

or as summarized in the rule above

$$(3 \times 10^3)^2 = 3^2 \times 10^{2 \times 3} = 9 \times 10^6.$$

Fractional powers, or roots. Taking a root of a number expressed in scientific notation is analogous to forming an integer power. The rule can be stated this way:

> *The multiplying factor in the result is the root of the multiplying factor in the original number; the exponent is obtained by <u>dividing</u> the original exponent by the degree (square, cube, etc.) of the root.*

This rule is really the same as that given for taking integer powers; however in the case of a root the power is a fraction rather than an integer. This can be seen in the following example:

Take the square root of 3.6×10^3.

DISCUSSION: Taking a square root is the same as raising to a power of ½. Thus
$$\sqrt{3.6 \times 10^3} = (3.6 \times 10^3)^{1/2} = (36 \times 10^2)^{1/2} = \sqrt{36} \times 10^{2/2} = 60.$$

Notice the trick in this example of making the power of ten an even power (10^2) so that it is easily raised to the ½-power.

Skill Drill 1

This drill deals with material of the preceding Review, "Dealing with Numbers." Go through the whole drill, building up speed and accuracy as you proceed. Space is provided for your work.

Periodically you should check your solutions against those given on the corresponding answer page, which is found on the reverse side of each sheet.

1. Express the following numbers in scientific notation, retaining three digits (3 significant figures) in the multiplying factor:

(a) 0.0344

(b) 25.5

(c) 89,310

(d) -735

(e) 0.000542

2. Express the following numbers as their decimal equivalents:

(a) 99.4×10^{-2}

(b) 4.60×10^{6}

(c) 0.4394×10^{3}

(d) -5.50×10^{-4}

(e) 1.41×10^{0}

3. For each of the following numerical expressions *first* make an order of magnitude estimate of the result, *then* find a rough approximation to one or two significant figures. Make the order of magnitude estimate entirely in your head; pencil and paper, but not a calculator, may be used for the rough approximation.

(a) (2.16)(8.151)

(b) (5239)(-6.9)

(c) 0.318/88

Skill Drill 1 — SOLUTIONS AND ANSWERS

1. Express the following numbers in scientific notation, retaining three digits (3 significant figures) in the multiplying factor:

(a) 0.0344 3.44×10^{-2}

(b) 25.5 2.55×10^{1}

(c) 89,310 8.93×10^{4}

(d) -735 -7.35×10^{2}

(e) 0.000542 5.42×10^{-4}

2. Express the following numbers as their decimal equivalents:

(a) 99.4×10^{-2} 0.994

(b) 4.60×10^{6} $4,600,000$

(c) 0.4394×10^{3} 439.4

(d) -5.50×10^{-4} -0.000550

(e) 1.41×10^{0} 1.41

3. For each of the following numerical expressions *first* make an order of magnitude estimate of the result, *then* find a rough approximation to one or two significant figures. Make the order of magnitude estimate entirely in your head; pencil and paper, but not a calculator, may be used for the rough approximation.

(a) (2.16)(8.151) $\sim (1)(10) = 10$

$\simeq (2)(8) = 16$

(b) (5239)(-6.9) $\sim (10^{3})(-10) = -10^{4}$

$\simeq (5 \times 10^{3})(-7) \simeq 4 \times 10^{4}$

(c) 0.318/88 $\sim 10^{-1}/10^{2} = 10^{-3}$

$\simeq 3 \times 10^{-1}/(9 \times 10^{1}) \simeq 3 \times 10^{-3}$

(d) 4.66/0.413

4. As in the previous problem first get an order of magnitude answer, then a rough approximation of each of the following expressions:

(a) (92.9)(0.666)/9.33

(b) $\dfrac{649}{(9.367)(-3438)}$

(c) (0.02250)(-8.872)(-70.6)

(d) $\dfrac{33.8}{(840)(-8.09)}$

5. Express each of the following roots or powers as a single number in scientific notation with three significant figures. For this problem use a calculator to take products or roots of the multiplying factors, but find the powers of ten mentally.

(a) $\sqrt{8.72 \times 10^2}$

(b) $\sqrt{46.7 \times 10^3}$

(c) $(3.47 \times 10^{-3})^2$

(d) $(-7.15 \times 10^3)^3$

6. Write each of the following expressions as a single number in scientific notation, rounding off the result to the correct number of significant figures.

(a) $6.56 \times 10^3 + 0.30 \times 10^2$

(b) $4480 - (6.75 \times 10^4)$

(d) 4.66/0.413 $\sim 1/10^{-1} = 10$

$$\simeq 4/(4 \times 10^{-1}) = 10$$

4. As in the previous problem first get an order of magnitude answer, then a rough approximation of each of the following expressions:

(a) (92.9)(0.666)/9.33 $\sim (10^2)(1)/10 = 10$

$$\simeq 10^2 (7 \times 10^{-1})/10 = 7$$

(b) $\dfrac{649}{(9.367)(-3438)}$ $\sim 10^2/(-10 \times 10^3) = -10^{-2}$

$$\simeq \frac{6 \times 10^2}{10 (-3 \times 10^3)} = -2 \times 10^{-2}$$

(c) (0.02250)(-8.872)(-70.6)

$$\sim 10^{-2}(-10)(-10^2) = 10$$

$$\simeq (2)(-9)(-7) \times 10^{-2+0+1} \simeq 14$$

(d) $\dfrac{33.8}{(840)(-8.09)}$ $\sim 10/(-10^3 \times 10) = -10^{-3}$

$$\simeq \frac{30}{(8)(-8)} \times 10^{-2} \simeq -0.5 \times 10^{-2}$$

5. Express each of the following roots or powers as a single number in scientific notation with three significant figures. For this problem use a calculator to take products or roots of the multiplying factors, but find the powers of ten mentally.

(a) $\sqrt{8.72 \times 10^2} = \sqrt{8.72} \times 10^1 = 2.95 \times 10^1 = 29.5$

(b) $\sqrt{46.7 \times 10^3} = \sqrt{4.67 \times 10^4} = \sqrt{4.67} \times 10^2 = 2.16 \times 10^2 = 216$

(c) $(3.47 \times 10^{-3})^2 = (3.47)^2 \times 10^{-6} = 12.0 \times 10^{-6}$

(d) $(-7.15 \times 10^3)^3 = (-7.15)^3 \times 10^9 = -366. \times 10^9 = -3.66 \times 10^{11}$

6. Write each of the following expressions as a single number in scientific notation, rounding off the result to the correct number of significant figures.

(a) $6.56 \times 10^3 + 0.30 \times 10^2 = (6.56 + 0.030) \times 10^3 = 6.59 \times 10^3$

(b) $4480 - (6.75 \times 10^4) = (0.448 - 6.75) \times 10^4 = -6.30 \times 10^4$

(c) $(3.82 \times 10^2) + 8454$

(d) $46.4 + 0.04$

7. In each of the following expressions rewrite each factor and term in scientific notation, then compute its final value expressed in scientific notation with three significant figures. Use a calculator only for combining multiplying factors, not for finding powers of ten.

(a) $(61.3)(0.00541)$

(b) $(27.5 + 0.669)$

(c) $\dfrac{10190}{(85.1 - 0.79)}$

(d) $\dfrac{2345}{(0.0783)(-9.46)}$

8. For each of the following expressions *first* make a rough estimate, and *then* compute a more precise value retaining the appropriate number of significant figures. Feel free to use any of the features of your calculator to get the final answer.

(a) $(8.12 \times 10^3)(24.72 \times 10^{-6})$

(b) $(9.8 \times 10^{-2})/(39.4 \times 10^4)$

(c) $(1.38 \times 10^3) + (7.0 \times 10^2)$

(d) $\sqrt{(59.1)(5.35)}$

(c) $(3.82 \times 10^2) + 8454 = (0.382 + 8.454) \times 10^3 = 8.836 \times 10^3$

(d) $46.4 + 0.04 = 4.64$ (The 0.04 term does not contribute significantly to the properly rounded final result.)

7. In each of the following expressions rewrite each factor and term in scientific notation, then compute its final value expressed in scientific notation with three significant figures. Use a calculator only for combining multiplying factors, not for finding powers of ten.

(a) $(61.3)(0.00541) = (6.13)(5.41 \times 10^{-3}) = 33.2 \times 10^{-2}$

(b) $(27.5 + 0.669) = (2.75 + 0.07) \times 10^1 = 2.82 \times 10^1$

(c) $\dfrac{10190}{(85.1 - 0.79)} = \dfrac{1.019 \times 10^3}{(8.51 - 0.079) \times 10^1} = 0.121 \times 10^2$

(d) $\dfrac{2345}{(0.0783)(-9.46)} = \dfrac{2.345 \times 10^3}{(7.83 \times 10^{-2})(-9.46)} = -3.17 \times 10^3$

8. For each of the following expressions *first* make a rough estimate, and *then* compute a more accurate value retaining the appropriate number of significant figures. Feel free to use any of the features of your calculator to get the final answer.

(a) $(8.12 \times 10^3)(24.72 \times 10^{-6}) \approx (8 \times 10^3)(2 \times 10^{-5}) = 16 \times 10^{-2} \approx 0.2$

$$= 0.201$$

(b) $(9.8 \times 10^{-2})/(39.4 \times 10^4) \approx 10^{-1}/(4 \times 10^5) = 0.25 \times 10^{-6} \approx 2 \times 10^{-7}$

$$= 2.49 \times 10^{-7}$$

(c) $(1.38 \times 10^3) + (7.0 \times 10^2) \approx (1.4 + 0.7) \times 10^3 \approx 2 \times 10^3$

$$= 2.10 \times 10^3 \text{ (The 2nd term contributes to the last significant place of the first term.)}$$

(d) $\sqrt{(59.1)(5.35)} \approx \sqrt{60(5)} = \sqrt{3 \times 10^2} \approx 20$

$$= 17.8$$

Review 2 — Units to Go with the Numbers

In the examples discussed in Review 1 no mention is made of the nature of the things to which the numbers refer. But in fact, throughout physics all numbers are associated with physical entities which, in principle, can be measured. The unit of measurement must always appear with the number. For example:

- The length of the San Francisco Bay Bridge is 1378 *meters*.
- The time for the moon to circle the Earth is 27.3 *days*.

Students should always keep in mind that the numbers used in physics almost always are accompanied by some unit of measurement, the most common of which are so-called metric units.

SYSTEMS OF UNITS

Metric units. You will come across many different units in your study of physics. Some will be familiar to you, such as inches, seconds, and grams. Some will be unfamiliar such as newtons, pascals, and hertz. However the principle units used in much of physics are those based on the following three "metric" quantities:

- the meter (m), a unit of length,
- the second (s), a unit of time, and
- the kilogram (kg), a unit of mass or "amount of matter."

These three basic units (together with a few others related to electricity and some other branches of physics) are the foundation of a system of interrelated units known as the Système International, or S.I. units. The English system of units, including quantities such as the foot and the pound, because of its value in everyday life in the U.S. and in some parts of engineering, is also used to some extent in introductory physics.

Prefixes. Since the numbers which describe physical quantities range from the very large to the very small, a set of standard prefixes are used to designate convenient-sized units which differ by multiples of ten from the primary units. A familiar example is the centimeter (cm) which is a hundredth of a meter. Some other common prefixes are listed in the following table:

Multiple	Prefix	Symbol	Example
10^6	mega-	M	Power: 1 million watts (10^6 W) is a mega-watt (1 MW).
10^3	kilo-	k	Mass: 1000 grams (10^3 g) is a kilogram (1 kg).
10^{-3}	milli-	m	Length: a thousandth of a meter (10^{-3} m) is a millimeter (1 mm).
10^{-6}	micro-	μ	Time: a millionth of a second (10^{-6} s) is a microsecond (1 μs).
10^{-9}	nano-	n	Length: a millionth of a millimeter (10^{-9} m) is a nanometer (1 nm).

SIZING UP AN ANSWER

Sizes of things. How big or small are some of these units? Just as numerical estimates are useful before doing a detailed calculation, it is worthwhile to have a sense for the physical size of the quantities involved in problems. Your ability to estimate the approximate size of things will develop with experience. But as a beginning, consider the following list of comparisons among some important units or familiar objects:

Length:
- 2½ cm is about an inch
- a meter is about 3 feet (length of a yardstick)
- a km is almost 2/3 mile
- a typical adult's arm span or height is almost 2 m

Mass and volume:
- a liter (1000 cm^3) is about a quart
- a liter of water has a kg of mass
- a kg of stuff weighs about 2½ pounds
- an average adult's mass is about 70 kg

Time:
- 10^5 seconds is somewhat more than a day
- a human heart beats about once per second
- it takes about 2½ seconds for an object to fall from the top of a 10 story building to the ground, or about ½ second from a table top to the floor.

Making estimates. Finding the exact equivalent of a measurement made in one set of units in terms of another set of units (conversion of units) will be discussed in a later section of this review. But it isn't necessary to go through a formal conversion process to estimate the size of something in familiar terms, as in the following example:

A runner covers 10^6 cm during a race . Is it likely that the runner is a sprinter?

DISCUSSION: The answer is NO. There are 100 cm in a meter so that 1,000,000 cm (10^6 cm) contains 10,000 meters or 10 km. The race is 7 to 8 miles — definitely not a sprint.

COMBINATIONS OF UNITS

An example: speed. Calculations in physics often require that we combine quantities having different units. The result is a quantity whose unit is a combination of the units which went into the calculation. As an example consider a determination of speed, the rate at which an object moves from one position to another:

What is the speed of an automobile which travels 300 meters in 60 seconds?

DISCUSSION: Speed, the rate of change of position, is the number of units of length moved divided by the time it takes. Using the symbol Δ, which can be read "change in" (or something similar), the required calculation is

$$Speed = \frac{change\ in\ distance}{change\ in\ time} = \frac{\Delta d}{\Delta t} = \frac{300\ m}{60\ s} = 5.0\ m/s$$

The unit m/s, which also may be written $m\cdot s^{-1}$, stands for "meters per second." It can be thought of as the number of meters moved per unit time.

Dimensionless ratios. In physics there is no meaning to an answer which is unaccompanied by units, *except* when the answer refers to the comparative size of two quantities having the same units. For example the number of times the diameter of a circle goes into its circumference is $\pi = 3.1416...$, a so-called "dimensionless" or unitless number.

Getting the units right. To be sure of the correct units in the results of a calculation, it is strongly recommended that units are included with all the numerical quantities you write down. Then

> • *the several units which appear in an expression multiply, divide, or cancel one another to form a resultant unit.*

As an example, consider the following problem:

An electron travels from the back end of a television picture tube to the screen, a distance of 50 cm, in 2.5 μs. What is its average speed?

DISCUSSION: It may prove convenient to change to other units during the calculation so that the final units will be more conventional. For instance

$$Speed = \frac{50\ cm}{2.5\ \mu s} = \frac{50\times 10^{-2} m}{2.5\times 10^{-6} s} = 20.\times 10^{4}\ m/s\ .$$

Combinations of units with special names. Some combinations of units are used so often in physics or in technical work that they are given special names. For instance, the term "knot" stands for nautical miles per hour. This unit is used in the following example, which shows how specially named composite units can be manipulated.

An aircraft flies in a straight line at 400 knots for 1.5 hours. How far does it travel?

DISCUSSION: By writing out the composite unit "knots" in terms of its component units, a cancellation of units occurs. The result is then purely in units of length, as follows:

Distance= (400 naut.mi/h̶r̶)(1.5 h̶r̶) = 600 naut.miles .

Many specially named units will crop up in your study of physics. It is worthwhile to know precisely how each of the more common ones are defined.

CONVERSION OF UNITS

Conversion factors. Knowing the value of a physical quantity in terms of some particular unit, you may frequently want to know its equivalent in terms of another unit. Change (or "conversion") of units can be made by multiplying (or dividing) the original unit by an appropriate "conversion factor" whose value can be found, if necessary, from information listed in a table.

A conversion factor is simply the ratio of a quantity stated in one unit to the same quantity stated in another unit. Thus both units will appear when the conversion factor is written out. For instance, suppose an answer stated in feet is to be converted to the answer stated in meters. A table might tell you "1 m = 3.28 ft;" appropriate conversion factors are either (3.28 ft/meter) or (1 m/3.28 ft).

Multiply or divide by the appropriate conversion factor so that the unwanted unit cancels, leaving the desired unit in the result.

This illustrated for a meters-to-feet conversion as follows:

How many feet does a runner go in a 200-meter dash?

DISCUSSION: Multiplication by the first of the above conversion factors results in a cancellation of "meters." Thus

Distance in feet = (200 m̸)(3.28 ft/m̸) = 656 ft.

Chains of conversions. Sometimes it is necessary to convert between two units for which conversion information cannot be found directly from a table. In such a case several known conversion factors can be used so that all the necessary cancellations of units take place. Here is an example, again involving length units:

How many meters are in exactly 1 mile? Use the following conversion information: 5280 feet are in a mile, 12 inches are in a foot, 2.54 cm are in an inch, and 100 cm are in a meter. DISCUSSION: Chain together the appropriate conversion factors as follows.

$$\text{Distance in meters} = (1 \text{ mi}) \left(\frac{5280 \text{ ft}}{1 \text{ mi}} \right) \left(\frac{12 \text{ in}}{1 \text{ ft}} \right) \left(\frac{2.54 \text{ cm}}{1 \text{ in}} \right) \left(\frac{1 \text{ m}}{100 \text{ cm}} \right)$$

$$= 1609 \text{ m.}$$

All the numerical conversion factors in this problem happen to be exact, as they are defined to have these values by international agreement. Therefore the result of the unit conversion is good to as many significant figures as is needed. (We chose to round off to four figures in this example.)

Unlike those in the above example, some conversion factors are determined from laboratory measurements. For these, the number of significant figures after conversion are limited by the number of significant figures in the conversion factors.

Skill Drill 2

These exercises should help you build up your speed and comfort level in using and converting units, and in making rough estimates of the sizes of things. Continue to use the calculating and estimating skills discussed in Review 1. Space is provided for your work. Frequently check yourself by turning over a page and looking at the solution.

1. Answer the following questions without looking up any conversion factors.

(a) How many mm in a km?

(b) How big is 1 μm (also called a "micron") in meters and, approximately, in inches?

(c) A mile is about 1000 paces (double steps). Roughly how many meters are in a pace?

(d) What is the size measured in angstroms (10^{-10} m) of a 500 nm wavelength of light?

2. Make an estimate (1 significant figure) of your age in seconds. Use information such as the number of seconds in an hour, hours in a day, etc.

3. Make an estimate of how many ordinary bricks would go into a brick wall 1 ft thick, 5 ft high, and 1 mile long.

Skill Drill 2 — SOLUTIONS AND ANSWERS

1. Answer the following questions without looking up any conversion factors.

(a) How many mm in a km?   *1000 mm in a meter, and 1000 m in a km; hence 1000 × 1000 mm (10⁶ mm) in a km.*

(b) How big is 1 μm (also called a "micron") in meters and, approximately, in inches?

$$1 \text{ micron} = 10^{-6} \text{ m}$$
$$1 \text{ inch} \simeq 2\frac{1}{2} \text{ cm} = 2\frac{1}{2} \div 100 \text{ m} = 0.025 \text{ m}; \text{ hence}$$
approx. $0.025 / 10^{-6} = 25,000 \text{ μm}$ in an inch.

(c) A mile is about 1000 paces (double steps). Roughly how many meters are in a pace?

$$1 \text{ pace} \simeq 10^{-3} \text{ mile and } 1 \text{ mi} \simeq 1\frac{1}{2} \text{ km} = 1500 \text{ m};$$
hence $1 \text{ pace} \simeq 10^{-3} \times 1500 = 1.5 \text{ m}$

(d) What is the size measured in angstroms (10^{-10} m) of a 500 nm wavelength of light?

$$500 \text{ nm} = 500 \times 10^{-9} \text{ m} = 5000 \times 10^{-10} \text{ m};$$
hence $500 \text{ nm} = 5000$ angstroms

2. Make an estimate (1 significant figure) of your age in seconds. Use information such as the number of seconds in an hour, hours in a day, etc.

$$\text{Each 10 years is about } (10 \times 400 \text{ days})$$
$$\simeq (4000 \times 25) \text{ hours} = 100 \times 10^3 \text{ hr}$$
$$\simeq (10^5 \times 4000) \text{ seconds} \simeq 4 \times 10^8 \text{ S}$$

If you are 25 years old you have lived approximately 1 billion (10^9) seconds.

3. Make an estimate of how many ordinary bricks would go into a brick wall 1 ft thick, 5 ft high, and 1 mile long.

$$\text{Volume of wall} \simeq 1 \times 5 \times 5000 = 25,000 \text{ cu. ft}$$
$$\text{Volume of a brick} \simeq 1 \times \frac{1}{2} \times \frac{1}{4} \simeq (\frac{1}{10}) \text{ cu. ft}$$
$$\text{Hence nb. of bricks} \simeq 25,000 / \frac{1}{10} \text{ bricks}$$
$$= 250,000 \text{ bricks}$$

4. If people formed a human chain by holding hands, roughly how many people would it take to span your state?

5. Think of how big a box you could just barely squeeze into, i.e., a box with approximately your body volume. (a) How many cubic cm (cm^3) does it contain? (b) How many liters? (c) How much water (in kg) would it contain?

6. Density is the amount of mass of a body divided by its volume. A liter of water has 1 kg of mass. What is the density of water in g/m^3?

7. Approximately how long would it take (in seconds) for a light beam (speed = 3×10^8 m/s) to cross an atomic nucleus (diameter approximately 10^{-15} m, also called a "femtometer")? (This may be about as small an interval of time as has any physical meaning.)

8. Calculate how much a U.S. dollar is worth in French francs using this information: \$1 = £0.529 (British pounds), £1 = 2.96 DM (German deutsche marks), and 1 DM = 3.35 F (French francs).

4. If people formed a human chain by holding hands, roughly how many people would it take to span your state?

$$For\ each\ 100\ miles = 250\ km - 250,000\ m,$$
$$it\ would\ require,\ at\ 2m\ per\ arm\ span,$$
$$approx.\ \frac{250,000}{2}\ people = 125,000\ people.$$

5. Think of how big a box you could just barely squeeze into, i.e., a box with approximately your body volume. (a) How many cubic cm (cm³) does it contain? (b) How many liters? (c) How much water (in kg) would it contain?

(a) Typical volume $\simeq (200\ cm)(30\ cm)(10\ cm) = 6 \times 10^4 cm^3$
(b) $(6 \times 10^4 cm^3)(1\ell / 10^3 cm^3) = 60\ liters$
(c) At 1 Kg per liter $\rightarrow$ 60 Kg (which is equivalent to about 130 lb, close to an adult body weight)

6. Density is the amount of mass of a body divided by its volume. A liter of water has 1 kg of mass. What is the density of water in g/m³?

$$1\ liter = 10^3 cm^3,\ and$$
$$1\ m^3 = (100\ cm)^3 = 10^6 cm^3.\ Hence$$
$$Density = \left(\frac{1\ Kg}{1\ell}\right)\left(\frac{10^3 g}{1Kg}\right)\left(\frac{1\ liter}{10^3 cm^3}\right)\left(\frac{10^6 cm^3}{1\ m^3}\right)$$
$$= 10^6 g/m^3$$

7. Approximately how long would it take (in seconds) for a light beam (speed = 3×10^8 m/s) to cross an atomic nucleus (diameter approximately 10^{-15} m, also called a "femtometer")? (This may be about as small an interval of time as has any physical meaning.)

$$Time = \frac{Distance}{speed} = \frac{10^{-15}\ m}{3 \times 10^8 m/s} \simeq 3 \times 10^{-24}\ s$$

8. Calculate how much a U.S. dollar is worth in French francs using this information: $1 = £0.529 (British pounds), £1 = 2.96 DM (German deutsche marks), and 1 DM = 3.35 F (French francs).

$$value\ of\ dollar = (\$1)\left(\frac{£0.529}{\$1}\right)\left(\frac{2.96\ DM}{£1}\right)\left(\frac{3.35\ F}{1\ DM}\right)$$
$$= 5.25\ F$$

9. How fast is a mi/hr in m/s? First make an estimate, then convert units using this information: 5280 ft in a mile, 3.28 ft in a meter, 60 s in a minute, and 60 min in an hour.

10. Express atmospheric pressure in newtons per square meter (N/m^2) given these conversions: atmospheric pressure = 14.7 lb/in^2, 2.54 cm = 1 inch, 1 N = 0.225 lb. (Note: 1 N/m^2 is called a pascal, Pa.)

11. You are going 50 km/hr in your car when you suddenly see a stalled car in your path. If the time it takes before you begin applying the brake pedal (reaction time) is 0.50 s, how far has your car gone during this interval in meters, and in feet? (3.28 ft per meter.)

12. A watt (W) is a unit of power (rate of energy usage). (a) How many 1 kW baseboard heaters can be run by the output of a 200 MW nuclear power plant?   (b) What is the equivalent horsepower? (1 hp = 746 W.)

9. How fast is a mi/hr in m/s? First make an estimate, then convert units using this information: 5280 ft in a mile, 3.28 ft in a meter, 60 s in a minute, and 60 min in an hour.

$$Roughly \quad 1\,mi/hr. \approx \frac{1500\,m}{3000\,s} = \frac{1}{2}\,m/s$$

$$More\ precisely: \left(\frac{1\,\cancel{mi}}{\cancel{hr}}\right)\left(\frac{5280\,ft}{1\,\cancel{mi}}\right)\left(\frac{1\,m}{3.28\,\cancel{ft}}\right)\left(\frac{1\,\cancel{hr}}{60\,\cancel{min}}\right)\left(\frac{1\,\cancel{min}}{60\,s}\right)$$

$$= 0.447\,m/s$$

10. Express atmospheric pressure in newtons per square meter (N/m²) given these conversions: atmospheric pressure = 14.7 lb/in², 2.54 cm = 1 inch, 1 N = 0.225 lb. (Note: 1 N/m² is called a pascal, Pa.)

$$Atmospheric \ \ pressure$$
$$= \left(\frac{14.7\,\cancel{lb}}{\cancel{in}^2}\right)\left(\frac{1\,\cancel{in}}{2.54\,\cancel{cm}}\right)^2 \left(\frac{1\,N}{0.225\,\cancel{lb}}\right)\left(\frac{100\,\cancel{cm}}{1\,m}\right)^2$$
$$= 1.013 \times 10^5 \, N/m^2 \approx 10^3 \, Pa$$

11. You are going 50 km/hr in your car when you suddenly see a stalled car in your path. If the time it takes before you begin applying the brake pedal (reaction time) is 0.50 s, how far has your car gone during this interval in meters, and in feet? (3.28 ft per meter.)

$$Distance = (speed)(time)$$
$$\left(\frac{50\,km}{hr}\right)(0.5s)\left(\frac{10^3 m}{km}\right)\left(\frac{1\,hr}{3600\,s}\right)$$
$$= 25\,m$$
$$= 25\,m\left(\frac{3.28\,ft}{1\,m}\right) = 82\,ft$$

12. A watt (W) is a unit of power (rate of energy usage). (a) How many 1 kW baseboard heaters can be run by the output of a 200 MW nuclear power plant? (b) What is the equivalent horsepower? (1 hp =746 W.)

$$(a) \ No.\ of\ heaters = \frac{200\,MW}{1\,kW} = \frac{200 \times 10^6\,W}{10^3\,W}$$
$$= 200,000\ heaters$$

$$(b) \ Power = 200 \times 10^6\,W\left(\frac{1\,hp}{746\,W}\right) = 270,000\,hp$$

Review 3 — Words, Symbols, and Pictures

Learning physics requires something other than committing to memory a large number of names, definitions, and laws. It requires using information in an active way and in a variety of contexts. The object is to develop, largely through problem solving, an effective understanding of relatively few concepts. Invariably the questions and problems used in this learning process are presented in words, accompanied occasionally by a clarifying picture. In lower grades, you may have come to know these as "story problems."

WORDS TO SYMBOLS — ALGEBRAIC STATEMENTS

Algebraic statements. Not all physics "story or word problems" need be put into mathematical form, but often this is the case. Two steps needed to translate word statements into mathematical (algebraic) statements are these:

> *(1) The important quantities in the problem are identified and assigned symbols, such as letters of the alphabet; and*
> *(2) the situation which is described in words is then interpreted in terms of mathematical relationships among these symbols.*

Consider the following non-physics statement as an example of this two step process:

In seven years Jack will be twice as old as Bob was five years ago.

DISCUSSION: (1) The statement has to do with ages, so appropriate symbols are assigned to represent these: J for Jack's <u>present</u> age, and B for Bob's. Sometimes these letters are called "algebraic variables." As they represent measurable physical quantities, they are associated with both a number and a unit (in this case years).

(2) The relationship among the variables demands a clear understanding of the meaning of the words: "will be" and "was" imply that we must set something having to do with a future age equal to something involving some past age, requiring addition and subtraction of numbers from the present ages, respectively; "twice as old" signifies multiplication by two. Thus as an algebraic statement we have:

$$J + 7 \text{ yr} = 2(B - 5 \text{ yr}).$$

Since it is meaningless to add or subtract unlike things, when choosing units it is important to keep in mind the following:

> *Every term in an algebraic expression must have the same units.*

Thus, in the last example, if J and B were chosen to represent ages in months rather than in years, the correct expression would be

$$J + 84 \text{ mo} = 2(B - 60 \text{ mo}).$$

Algebraic quantities. In choosing symbols to represent physical quantities it is convenient to choose letters which are abbreviations of the names they represent, although particular symbols are traditionally associated with certain types of quantities. For instance the symbol x often represents the distance along a horizontal line from some origin. Subscripts are frequently used; for example x_1 can be used to denote the position of an object at one time and x_2 to denote its position at a later time.

Physics also makes use of the Greek alphabet. It is worthwhile being familiar with some of the more commonly used Greek letters, as given in the following table:

Some Greek letters often used in beginning physics.*					
α	alpha	θ	theta	σ	sigma
β	beta	λ	lambda	Σ	UC sigma
γ	gamma	μ	mu	φ	phi
δ	delta	υ	nu	ω	omega
Δ	UC delta	ρ	rho	Ω	UC omega

*Lower case unless indicated UC.

WORDS TO SYMBOLS — MAKING AND USING DIAGRAMS

Representing relationships — especially spatial relationships. To express the connections among algebraic quantities, a limited number of symbols can be used to represent an almost unlimited number of word statements. For instance, the expression "added to" usually appears in an algebraic statement in terms of the symbol "+", but this same symbol can be used to denote such words as "increased," "enlarged," "grow," "taller," "more," etc.

Practicing with word problems is invaluable for developing skill at symbolizing statements. However, in many physics situations, especially where a problem has to do with relative positions of objects, drawing a simple diagram makes it easier to grasp the situation, so that the words can be readily interpreted.

A simple drawing which depicts the spatial arrangements described in a problem greatly aids understanding.

This is illustrated by the next example:

A radar operator on a small island (labelled I below) determines that a ship is 15 km due north of him. Some time later the ship is 10 km southeast of the island. If the ship moves in a straight line, draw a diagram showing the ship's path.

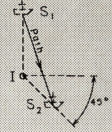

DISCUSSION: Many problems in physics involve the relative orientation of objects in a plane, such as the situation described here. Appropriate drawings are much like maps. It is often useful to label the angles which show the orientation of important lines in the diagram, as in this sketch showing the earlier and later positions of the ship S_1 and S_2.

A drawing need not be fancy; a reasonably careful freehand sketch will usually do.

Getting an answer from a diagram. Not only can a simple sketch help you to understand a problem, but it may sometimes provide a quick insight into the answer, as in the following example:

Carol is 6 inches taller and John is 3 inches taller than Marty. If Alfred is one inch shorter than Carol, how much taller (or shorter) is Alfred than John?

DISCUSSION: A sketch resembling height marks on a wall proves useful in this problem, as it reveals the answer essentially by inspection.

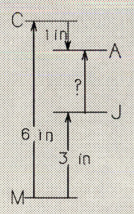

From the drawing we see directly that Alfred is 2 inches taller than John.

Although algebra can be used to solve a problem of this type, it isn't always necessary. In this example making the diagram leads to the answer without any complicated mathematics.

A diagram need not always represent a spatial relationship. It can be more abstract. For instance, if the above problem were changed to an "age" problem, by substituting "years" for "inches" and "older/younger" for "taller/shorter," the solution could be found using exactly the same diagram.

Skill Drill 3

This drill provides practice in translating word statements into equations and diagrams. Skillful reading is as important in problem solving as mathematical manipulation. Frequently check your work against the suggested solutions.

1. Translate the following English sentences into algebraic statements (equations), using symbols of your choosing where necessary. You do not need to algebraically simplify the equations.

(a) Charles is 18 years older than Morgan.

(b) My uncle is twice as old as your brother.

(c) John is 10 years older than Moe was 3 years ago.

(d) Ten *months* ago, Maggie was 18 years old.

(e) A British pound is three times as valuable as a German deutschemark.

(f) The temperature expressed in Kelvin is 273.1 degrees greater than the temperature expressed in degrees Celcius.

(g) The average x_{av} of x_1, x_2, and x_3 is the sum of these quantities divided by three.

(h) The "root mean square" average x_{rms} of x_1, x_2, and x_3 is the square root of one-third of the sum of the squares of these quantities.

Skill Drill 3 — SOLUTIONS AND ANSWERS

1. Translate the following English sentences into algebraic statements (equations), using symbols of your choosing where necessary. You do not need to algebraically simplify the equations.

(a) Charles is 18 years older than Morgan.

$$C = M + 18 \text{ yr}$$

(b) My uncle is twice as old as your brother.

$$U = 2B$$

(c) John is 10 years older than Moe was 3 years ago.

$$J = (M - 3\text{yr}) + 10 \text{ yr} = M + 7\text{yr}$$

(d) Ten *months* ago, Maggie was 18 years old.

$$M - \frac{10}{12} \text{ yr} = 18 \text{ yr}$$

(e) A British pound is three times as valuable as a German deutschemark.

$$P = 3D$$

(f) The temperature expressed in Kelvin is 273.1 degrees greater than the temperature expressed in degrees Celcius.

$$K = C + 273.1°$$

(g) The average x_{av} of x_1, x_2, and x_3 is the sum of these quantities divided by three.

$$X_{av} = \frac{X_1 + X_2 + X_3}{3}$$

(h) The "root mean square" average x_{rms} of x_1, x_2, and x_3 is the square root of one-third of the sum of the squares of these quantities.

$$X_{rms} = \sqrt{\frac{X_1^2 + X_2^2 + X_3^2}{3}}$$

(i) The density ρ of an object is defined as the mass of the object divided by the volume.

      2. The airport bus takes in as fares an average of D dollars per trip. It carries two children for every three adults, and carries an average of N adults per trip. The adult ticket fare is A dollars and the child's fare is half that. Write an equation which relates D, N, and A.

      3. With B buses per hour each with 50 passengers, the bus lane carries twice as many commuters per hour into the city as do the car lanes with C cars per hour with an average of 3 passengers per car. Relate B and C through an equation.

      4. Solve each of these problems by representing the data as a diagram:

(a) Josephine is shorter than Mary but taller than Gert. Who is tallest and who is shortest?

(b) If I were as rich as Carl I'd be twice as rich as Bob and three times as rich as John. Who is richer, Bob or John? How much richer?

(c) In language school for diplomats the most difficult languages require the longest period of training. French requires one-third as much time as Arabic, while Spanish requires one-third as much time as German, which in turn requires half as much time as Arabic. Which requires more training, French or German? French or Spanish?

(i) The density ρ of an object is defined as the mass of the object divided by the volume.

$$\rho = M/V$$

2. The airport bus takes in as fares an average of D dollars per trip. It carries two children for every three adults, and carries an average of N adults per trip. The adult ticket fare is A dollars and the child's fare is half that. Write an equation which relates D, N, and A.

No. of children/trip $= 2N/3$ and child's fare $= \dfrac{A}{2}$.

Thus, $D = NA + \dfrac{2N}{3}\dfrac{A}{2} = NA + \dfrac{NA}{3} = \dfrac{4}{3}NA$

3. With B buses per hour each with 50 passengers, the bus lane carries twice as many commuters per hour into the city as do the car lanes with C cars per hour with an average of 3 passengers per car. Relate B and C through an equation.

$$50B = 2(3C)$$
or
$$50B = 6C$$

4. Solve each of these problems by representing the data as a diagram:

(a) Josephine is shorter than Mary but taller than Gert. Who is tallest and who is shortest?

————— M
————— J
_____ G

Mary is tallest; Gert is shortest.

(b) If I were as rich as Carl I'd be twice as rich as Bob and three times as rich as John. Who is richer, Bob or John? How much richer?

Me ——————— C
B ————— ½C
J ————— ⅓C

Bob is richer by a factor of $\dfrac{1/2}{1/3} = 1\frac{1}{2}$

(c) In language school for diplomats the most difficult languages require the longest period of training. French requires one-third as much time as Arabic, while Spanish requires one-third as much time as German, which in turn requires half as much time as Arabic. Which requires more training, French or German? -French or Spanish?

A ————————— A
————— G = ½A
————— F = ⅓A
————— S = ⅓G

German is harder than French. French is harder than Spanish.

5. In a certain part of town numbered avenues run north-south whereas lettered streets run east-west. Use a "map-like"diagram to answer the following questions:

(a) Broadway crosses A-Street at a right angle. Is Broadway perpendicular or parallel to 3rd Avenue?

(b) Joe leaves his apartment at 4th Avenue and F-Street going north two blocks to H-Street and then east three blocks to 7th Avenue. He then goes south five blocks and west six blocks. Near what intersection does he end up?

(c) In what general direction does his apartment lie — NE, NW, SW, or SE?

6. An airplane sets out on a course which is due north and flies 100 miles. It then makes a left turn to the northwest and flies another 100 miles.

(a) Make an appropriate sketch.

(b) To go directly to its starting point should it make a left or right turn?

7. Draw a very simple sketch to illustrate each of the following situations:

(a) A ladder leaning against a wall (side view).

(b) A box resting on an incline.

(c) A ball rolling down an incline.

5. In a certain part of town numbered avenues run north-south whereas lettered streets run east-west. Use a "map-like" diagram to answer the following questions:

(a) Broadway crosses A-Street at a right angle. Is Broadway perpendicular or parallel to 3rd Avenue?

Parallel

(b) Joe leaves his apartment at 4th Avenue and F-Street going north 2 blocks to H-Street and then east three blocks to 7th Avenue. He then goes south 5 blocks and west 6 blocks. Near what intersection does he end up?

1st Avenue and C Street

(c) In what general direction does his apartment lie — NE, NW, SW, or SE?

NE

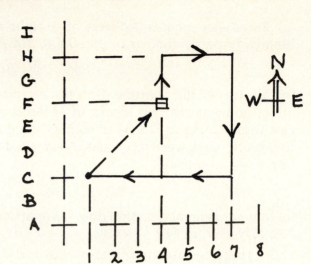

6. An airplane sets out on a course which is due north and flies 100 miles. It then makes a left turn to the northwest and flies another 100 miles.

(a) Make an appropriate sketch.

(b) To go directly to its starting point should it make a left or right turn?

Left

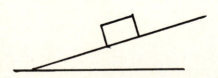

7. Draw a very simple sketch to illustrate each of the following situations:

(a) A ladder leaning against a wall (side view).

(b) A box resting on an incline.

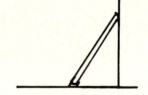

(c) A ball rolling down an incline.

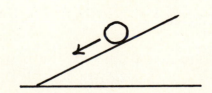

First Round Posttest — Optimum test time: 25 minutes or less

This is a brief test to give you a chance to see how much you have improved your facility with First Round skills since taking the Pretest. Some questions require a small amount of factual information (given in the Reviews) with which you may not have been familiar at the time you took the Pretest.

Use a calculator when needed, but not for order of magnitude calculations. Space is provided for answers but you will need a sheet of paper for most of the work. Keep track of your time and check your answers against those given at the end of this Round.

STARTING TIME_____

ANSWERS

1. For the following numbers give the order of magnitude *and* the value expressed to three significant figures in scientific notation.

(a) 0.0006524

_____

(b) $(8.221)^2$

_____

2. For the following numerical expressions first make an estimate (within a factor of two), then calculate an answer accurate to the appropriate number of significant figures and expressed in scientific notation.

(a) $\dfrac{(3439 \times 10^{-2} + 1.16)}{(0.172 \times 10^2)(73.1)}$

_____

(b) $\dfrac{\sqrt{0.859 \times 10^3}}{0.285 \times 10^3}$

_____

3. In term of S.I. units (meters/kilograms/seconds) give values of the following quantities. Give the answers in decimal form.

(a) 4000 µs

_____

(b) 40 miles per hour (approximately)

_____

(c) 100 grams

_____

4. Give a rough estimate (one or two significant figures) of the number of people in a well filled movie theater 50 feet by 100 feet in size.

_____

5. How many feet are there in 3.0 meters? Use this conversion information: 1 m = 39.37 in, 12 in = 1 ft.

_____

6. Express the following statements as algebraic equations using the indicated symbols:

(a) By selling 20 tickets at the standard admission price S, 10 tickets at the balcony price B, and 10 tickets at the children's price C, a total of T dollars was collected.

_____

(b) During her warmup Jenny covered a total distance D by running at her regular jogging speed V for a period of time T, then going half as fast for three times as long.

_____

7. Make a sketch which illustrates the following situation:

A fireman (F) carefully carried the baby down the 50 ft ladder which was set 20 feet out from the building wall. When he was halfway down he was directly above the mother (M).

8. Use a sketch or sketches to decide upon the answer to the following problem:

Centerville and Davenport are connected by a pair of straight sections of railroad track. Train A starts out at a steady speed from Centerville heading for Davenport at the same time that train B starts out from Davenport heading for Centerville. They pass each other one-third the distance from Centerville. When train B reaches Centerville, train A is

(a) halfway to Davenport
(b) two-thirds the distance to Davenport
(c) at Davenport.

_____

ENDING TIME_____

ANSWERS:

1. (a) 10^{-3}, 6.52×10^{-4}
   (b) 10^2, 6.76×10^1

2. (a) 0.03, 2.83×10^{-2}
   (b) 0.1, 1.03×10^{-1}

3. (a) 0.004 s (b) 16 m/s
   (c) 0.1 kg

4. 350 to 450 people

5. 9.8 ft

6. (a) 20S+10B+10C=T
   (b) D=VT+3VT/2

7.

8. (a)

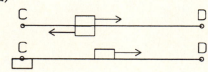

Essay: Why Is Physics so "Mathematical"?

One should make no mistake about it; physics is *not* mathematics. In fact, you will discover that some of the most challenging questions you will be asked in a physics course may require little if any mathematics to answer. Whereas mathematics is the study of abstract numbers and space and is ultimately based on the rules of counting and logic, physics is the study of the behavior of real, measurable objects and events. It is ultimately based on observations of the world around us. Despite these contrasts, the striking fact is that the laws of physics invariably present themselves in mathematical form. Isaac Newton did not simply find that the planets and the sun pull on one another. He found that the forces they exert on one another vary inversely as the square of the distance between them— which is certainly a mathematical statement. In fact, throughout physics we find that the basic rules governing the behavior of nature are readily expressed in mathematical form.

Why is there this curious connection between mathematics and physics? Actually nobody really knows. But the connection between the physical world and the mathematical world has a long history. In ancient Greece a prominent school of thought founded by Pythagoras held that all the secrets of nature could be explained in terms of the properties of pure numbers and geometry. The Pythagoreans even experimented in what we would call the physics of music, discovering that pleasing (or "harmonious") chords are produced by strings or pipes whose lengths are in proportion to certain whole numbers. They elevated their mystical attachment to numbers to virtually a religious level, and while doing so made a great deal of progress in pure mathematics. However, their contributions to the understanding of the physical world was, in fact, very limited.

At the beginning of the 17th century a new and clearer relationship between science and mathematics arose. At that time Johannes Kepler who, curiously, was driven by a compulsion to discover Pythagorean "harmonies" in the heavens, conceived the mathematical laws that describe the motion of the planets about the sun. He thereby laid important groundwork for physics. But it is to the work of Kepler's Italian contemporary Galileo Galilei that we usually look for the beginnings of modern science. Of special significance is his work on the motion of falling bodies, in which he uses mathematical reasoning to deduce how much distance an object dropping towards the ground might be expected to move in successive intervals of time. But Galileo went beyond mathematics by developing and conducting experiments to verify his predictions. He discovered that the distances fallen in 1, 2, 3, etc. units of time are proportional to 1^2, 2^2, 3^2, etc. units of length. This approach to discovery has set a precedent followed ever since: the basic laws of nature are stated in (usually simple) mathematical form, which enables us to make predictions and test them by experiment.

Yet the mystery remains as to why the basic laws are both mathematical and simple. In this respect Nature has been very kind; she has allowed us, through straightforward mathematics and careful measurement, to develop an elegant method to explore the world.

Round II — Nature at Its Simplest: Linear Behavior

The essence of physics has to do with describing and verifying the relationships among measured quantities. The mathematical expressions which are used to describe relationships in physics may sometimes be complicated. But, in a surprising number of cases, phenomena can be described by the very simplest of equations. This is the linear equation in which all variables occur to the first power. In this Round the algebra and graphing skills necessary to deal with linear physics are reviewed. The following pretest should help you assess your readiness to deal with such material.

_____

PRETEST — Optimum test time: 20 minutes or less.

Keep track of your starting and ending times. Put your answers in the space provided, then check them against those given following the test.

STARTING TIME_____

ANSWERS

1. If y is directly proportional to x, and x = 5 when y = 15, what is y when x = 6?

_____

2. Express the following statement as an equation: The cost C of buying granola in bulk is directly proportional to the weight purchased W. The constant of proportionality is the price per pound P.

_____

3. What is the slope and the y-intercept of the line representing the equation y = 2x − 3 ? Use your results to plot the line on this graph.

_____

_____

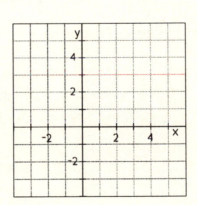

4. Extract a common factor from the following expression, and rewrite the expression with the factor separated from the remaining terms:

$$4x^3y - 2y^2x^2 + 4y^3x .$$

_____

5. Write the following expression as a polynomial in x:

$$(ax^2 - bx)(3 + 2x) .$$

_____

6. Find the solution of the following equation:

$$5(x+1) - 3(x+2) = 16x - 29 .$$

_____

7. Set up an equation and use it to solve the following problem:

A basket of 30 apples is to be divided into two portions, such that
2¼ times the smaller share is 15 greater than 2/3 of the other. How
large is the smaller share?

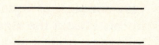

8. Solve the following pair of equations for x and y:

$$x + 2y = 5$$
$$y = 5 - 2x$$

9. Verify your solution to the previous
problem by drawing a pair of lines representing the equations on this graph.

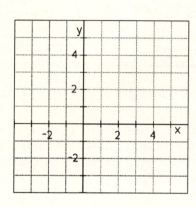

ENDING TIME_____

ANSWERS:

1. $y = 18$

2. $C = PW$

3. slope = 2

   intercept = −3

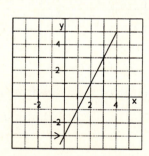

4. $\underline{2xy} \,(2x^2 - yx + 2y^2)$

5. $-3bx + (3a - 2b)x^2 + 2ax^3$

6. $x = 2$

7. $9S/4 = 15 + (2/3)(30 - S)$

   $S = 12$

8. $x = y = 5/3$

9.

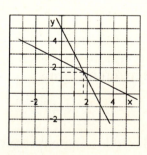

Review 4 — Proportion and Ratio

DIRECT PROPORTION

Definition and an example. Two quantities are said to be directly proportional to one another if an increase or a decrease of one of them by a certain factor is accompanied by an increase or decrease of the other one by the same factor. For example, if x is directly proportional to y, when x becomes twice as large, y also becomes twice as large. In symbols, direct proportionality between x and y can be written

$$y \propto x .$$

Physics abounds in phenomena for which measurable quantities are related in this way. A good example is the stretching of an ordinary helical spring, as pictured here. Suppose weights are to be suspended from such a spring. If the spring has a certain length to begin with, as weights are attached to the bottom of the spring the length will increase. Calling the amount of weight W (measured, for example, in pounds), and the corresponding increase in length x (measured, for example, in inches), for

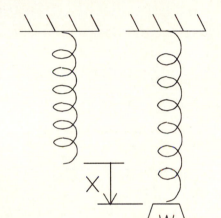

most springs it is found that W and x are directly proportional to one another. This is expressed

$$W \propto x .$$

Springs for which this relationship hold true are sometimes called "linear springs."

Proportionality constant. In every case of direct proportion, the relationship (as, for instance, between x and y) can also be described by an equation which looks like this:

$$y = kx,$$

where k is a quantity which depends neither on y nor x. k is called the "constant of proportionality."

Thus, the example given above of a spring and weight could have as well been described by the equation W = kx, where the proportionality constant k in this case is called the "spring constant." k is a quantitative measure of just how "stiff" or how "soft" the spring is.

A constant of proportionality must have units consistent with the other units in the proportionality equation. For instance, if the stretch of a spring is given in inches and the weight in pounds, the spring constant has units of pounds per inch (lb/in). This is illustrated in the following example problem:

One end of a spring is held by a support and a fish is attached to the lower end. The spring constant is $5.0 \cdot 10^{-3}$N/m. What is the weight of the fish if the spring is stretched out an additional 3.0 cm from its normal length? (N stands for newtons, the S.I. unit of force; weight is merely gravitational force.)

DISCUSSION: Assuming the spring to be "linear" we use the expression W = kx, where the proportionality constant k is the spring constant. In order to get W in appropriate units we must make a unit conversion (m to cm). Thus

$$W = kx = (5.0 \cdot 10^{-3} \text{N/m})(3.0 \text{ cm})$$
$$= (5.0 \cdot 10^{-3} \text{N/m})(3.0 \cdot 10^{-2} \text{m})$$
$$= 1.5 \text{ N}.$$

REPRESENTING DIRECT PROPORTION GRAPHICALLY

A straight line graph. Any equation relating quantities x and y corresponds to a curve drawn in a plane containing x and y axes (the "x-y plane").

> *When the relationship is one of direct proportion the curve is always <u>a straight line passing through the origin</u>.*

Moving from point to point along this line a change of x by some factor (for instance a doubling of x) corresponds to a change of y by the same factor (a doubling of y).

Dependent and independent variables. The graph shown here illustrates a convention which is often used in plotting the equations of physics. The value of one of the variable quantities (the "dependent variable," in this case y) is thought of as being set by whatever values are assigned to the other variable quantities in the equation (the "independent variables," in this case just x). Usually the dependent variable is plotted along the vertical axis, while independent variables (one for each curve) are plotted along the horizontal axis.

Slope and proportionality constant. Changes in the value of an independent variable give rise to changes in the value of the dependent variable. An important fact about these changes is the following:

> *For a proportional relationship such as y = kx the <u>changes</u> in the variables are also proportional.*

In the graph of the relationship y = kx, shown at the top of the next page, the change from one value of x to another has been labelled Δx, often called the "run." The corresponding change in y is labelled Δy, often called the "rise." It turns out that the constant of proportionality relating the rise and the run is also k. Thus we can write:

$$\Delta y = k \, \Delta x .$$

A measure of the steepness of the straight line is the ratio

$$\Delta y / \Delta x = k.$$

This ratio is called the "slope" of the straight line.

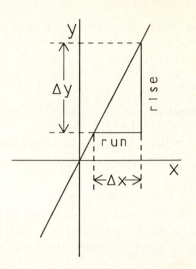

> *In the graph representing a proportional relationship the proportionality constant is equal to the slope of the line.*

Slope may be negative as well as positive; in this Review we will consider only positive slopes.

Again consider a linear spring, as in this example:

Below is a table giving data on the stretch x of a spring obtained when various weights W are suspended from it.

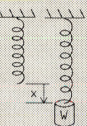

W(lb)	x(in)
0	0
1.0	0.50
3.0	1.49
5.0	2.51
10.0	4.99

Plot the points corresponding to each pair of values (W,x). Draw a straight line which fits the data and determine the spring constant by evaluating the slope of this line.

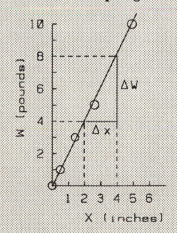

DISCUSSION: The deflection of spring is often used to determine force. Thus it is usual to consider x to be the independent variable and plot it on the horizontal axis. Set up appropriate scales on both axes and with a ruler draw a straight line which goes through the origin (0,0) and passes as close as possible to all the points. Then mark off with the ruler a convenient size run Δx (here 2.0 in has been chosen); next mark off the corresponding ΔW. The determination of k is as follows:

$$k = \frac{rise}{run} = \frac{\Delta W}{\Delta x} = \frac{4.0 \text{ lb}}{2.0 \text{ in}} = 2.0 \text{ lb/in} .$$

SOLVING PROBLEMS USING RATIOS

Setting up a ratio equation. Very often it isn't necessary to find the constant of proportionality in order to solve problems involving proportional relationships. When variable y changes, variable x changes by the same factor no matter what the proportionality constant may be. Thus if we know the ratio of a pair of independent variables we also know the ratio of the corresponding pair of dependent variables.

Labelling corresponding values of x and y by the same subscripts this relationship between ratios can be stated this way:

$$y_1/y_2 = [\text{the same factor}] = x_1/x_2 .$$

Sometimes you will see this written $y_1:y_2 = x_1:x_2$, but the ratio equation above is more useful.

>*For a proportional relationship the ratio of a pair of values of the dependent variable is <u>equal</u> to the ratio of a corresponding pair of values of the independent variable.*

Using ratio equations. In many problems in which $y \propto x$, three of the four quantities x_1, x_2, y_1, and y_2 are known, and we are asked to find the remaining quantity. A simple algebraic manipulation enables us to write an equation in which the unknown quantity is written in terms of the known quantities. We start by writing down the ratio expression

$$\frac{x_1}{x_2} = \frac{y_1}{y_2}.$$

Suppose x_1 is the quantity to be determined. The desired expression is obtained by multiplying both sides of the ratio equation by x_2. Since the x_2's on the left hand side cancel we are left with

$$x_1 = x_2\frac{y_1}{y_2} .$$

This way of rearranging an equation is also called "cross multiplication" since a factor (x_2) from one side of the equation is transferred diagonally across the equal sign. (See Review 6.) This manipulation is used in the following example:

Suppose we hang a 2.0 lb fish from the lower end of a linear spring causing it to deflect 2.5 cm. A larger fish causes a deflection of 3.0 cm. How much does the larger fish weigh?

DISCUSSION: Using subscripted x's for the two deflections and subscripted W's for the corresponding weights,
ratios can be set equal as follows:
$$x_2/x_1 = W_2/W_1 .$$
Thus
$$W_2 = W_1 (x_2/x_1)$$

$$= (2.0 \text{ lb})(3.0 \text{ cm}/2.5 \text{ cm}) = 2.4 \text{ lb} .$$

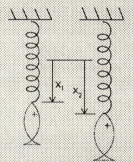

(Using a calibrated scale to mark off the deflections of a spring directly in force units is the principle of the spring balance or "fish scale.")

Sometimes it is useful to start the solution of a ratio problem by writing

$$\frac{x_1}{y_1} = \frac{x_2}{y_2} ,$$

a form of the ratio equation which is easily obtained by applying cross multiplication to the expressions given above.

Skill Drill 4

This drill emphasizes working with direct proportion, including finding proportionality constants and slopes of lines, and using ratios. Except for the first exercise these questions also provide more practice in handling units and doing word problems. As before, frequently check your answers.

1. This exercise is to remind you of the key points in Review 4.

(a) Express the proportionality $v \propto w$ as an equation in terms of a proportionality constant C. Let v be the dependent variable.

(b) Referring to part (a), if w = 2 inches when v = 4 pounds, what is the proportionality constant C? (Include units.)

(c) Refer to the graph in the first example in the Review: For a "rise" of 5.0 lb, what is the corresponding run?

(d) Consider two equal ratios x:3 = 3:5. Determine x.

2. Express each of the following statements as an equation which includes the given constant of proportionality. Make sure units are consistent:

(a) The rate of heat transferred W (in watts) across a section of window pane is directly proportional to the temperature difference ΔT across the glass (in degrees Celsius) The proportionality constant (κA) has units of watts/°C.

(b) The pressure p at the bottom of a lake and the depth of the lake d are related as a direct proportion. The proportionality constant is the weight of water per unit volume (usually written ρg). (NOTE: pressure units are always force units divided by area units, such as pounds/inch^2.)

3. A conversion factor C is a type of proportionality constant relating different units which are directly proportional to one another. Find the conversion factor which relates each of the following pairs of units, using the given information.

(a) Feet – centimeters: a 3.00 ft yardstick measures 91.4 cm.

(b) Barrels – gallons: a 1500 gal tank truck holds 47.6 bbl.

Skill Drill 4 — SOLUTIONS AND ANSWERS.

1. This exercise is to remind you of the key points in Review 4.

(a) Express the proportionality v ∝ w as an equation in terms of a proportionality constant C. Let v be the dependent variable.

$$v = Cw$$

(b) Referring to part (a), if w = 2 inches when   v = 4 pounds, what is the proportionality constant C? (Include units.)

$$C = \frac{4\ lb}{2\ in} = 2\ lb/in$$

(c) Refer to the graph in the first example in the Review: For a "rise" of 5.0 lb, what is the corresponding run?

Read from graph:
2.5 in

(d) Consider two equal ratios x:3 = 3:5.
Determine x.

$$\frac{x}{3} = \frac{3}{5}$$

$$\rightarrow x = 3\left(\frac{3}{5}\right) = \frac{9}{5}$$

2. Express each of the following statements as an equation which includes the given constant of proportionality Make sure units are consistent:

(a) The rate of heat transferred W (in watts) across a section of window pane is directly proportional to the temperature difference ΔT across the glass (in degrees Celsius) The proportionality constant (κA) has units of watts/°C.

$$W = kA\ \Delta T$$

$$units: \frac{watts}{°C}\ °C$$

$$= watts$$

(b) The pressure p at the bottom of a lake and the depth of the lake d are related as a direct proportion. The proportionality constant is the weight of water per unit volume (usually written ρg). (NOTE: pressure units are always force units divided by area units, such as pounds/inch².)

$$p = \rho g d$$

possible units:

$$\frac{lb}{in^3} \cdot in = \frac{lb}{in^2}$$

3. A conversion factor C is a type of proportionality constant relating different units which are directly proportional to one another. Find the conversion factor which relates each of the following pairs of units, using the given information.

(a) Feet – centimeters: a 3.00 ft yardstick measures 91.4 cm.

$$91.4\ cm = C(3.00\ ft)$$

$$C = \frac{91.4\ cm}{3.00\ ft} = 30.5\ cm/ft$$

(b) Barrels – gallons: a 1500 gal tank truck holds 47.6 bbl.

$$1500\ gal = C(47.6\ bbl)$$

$$C = \frac{1500\ gal}{47.6\ bbl} = 31.5\ gal/bbl$$

(c) Pressure (in pascals, Pa) – pressure (in pounds per square inch, psi): atmospheric pressure is 1.013×10^5 Pa and is also 14.7 psi.

4. The graph at the right plots the energy J (in joules) used by a light bulb versus the length of time t (in seconds) it is in use. Express the direct proportionality between J and t as an equation in terms of a proportionality constant W (in joules/second, also called watts). From the slope of the straight line in the graph determine W.

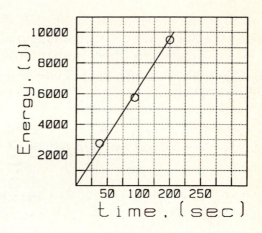

5. It takes more force F to push a heavy box along the floor than to push a light one. In fact, the force required is directly proportional to the weight of the box W. In other words $F = \mu W$; the proportionality constant μ is called the "coefficient of friction."

(a) Suppose you have to push with a force of 60 pounds to slide a filled packing crate weighing 300 pounds down the hall. What is the coefficient of friction?

(b) How much force is required to push it back up the hall after 200 pounds of equipment has been removed from the crate?

6. Use ratios to solve the following problems:

(a) A battleship sails 50 miles in the same time it takes a destroyer to sail 75 miles. The two ships set out to cross the Atlantic Ocean. After the destroyer has sailed 2500 mi, how far will the battleship have gone?

(b) On a car trip known from a very accurate map to be 300 miles, your odometer registers 270 miles. For each actual mile travelled how much does the odometer reading change?

(c) It takes 4 measures of ground coffee to make 7 cups of drink. How many measures should be used for 42 cups?

(c) Pressure (in pascals, Pa) – pressure (in pounds per square inch, psi): atmospheric pressure is 1.013×10^5 Pa and is also 14.7 psi.

$$1.013 \times 10^5 Pa = C \ 14.7 \ psi$$

$$C = \frac{1.013 \times 10^5 Pa}{14.7 \ psi} = 6.89 \ \frac{Pa}{psi}$$

4. The graph at the right plots the energy J (in Joules) used by a light bulb versus the length of time t (in seconds) it is in use. Express the direct proportionality between J and t as an equation in terms of a proportionality constant W (in Joules/second, also called Watts). From the slope of the straight line in the graph determine W.

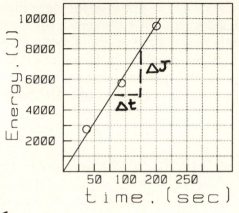

$$J = Wt$$

$$W = slope = \frac{\Delta J}{\Delta t}$$

$$= \frac{3000 \ J}{50 \ s}$$

$$= 60 \ J/s = 60 \ watts$$

5. It takes more force F to push a heavy box along the floor than to push a light one. In fact, the force required is directly proportional to the weight of the box W. In other words $F = \mu W$; the proportionality constant μ is called the "coefficient of friction."

(a) Suppose you have to push with a force of 60 pounds to slide a filled packing crate weighing 300 pounds down the hall. What is the coefficient of friction?

(b) How much force is required to push it back up the hall after 200 pounds of equipment has been removed from the crate?

(a) $F \longrightarrow \boxed{W}$

$$F = \mu W$$
$$\mu = 60 \ lb / 300 \ lb$$
$$= 0.20$$

(b) $F = 0.20 \ (100 \ lb)$
$$= 20 \ lb$$

6. Use ratios to solve the following problems:

(a) A battleship sails 50 miles in the same time it takes a destroyer to sail 75 miles. The two ships set out to cross the Atlantic Ocean. After the destroyer has sailed 2500 mi, how far will the battleship have gone?

$$\frac{X_B}{X_D} = \frac{50 \ mi}{75 \ mi} = 0.67$$

$$X_B = (2500 \ mi)(0.67)$$
$$= 1700 \ mi$$

(b) On a car trip known from a very accurate map to be 300 miles, your odometer registers 270 miles. For each actual mile travelled how much does the odometer reading change?

$$\frac{X_{act}}{X_{odom}} = \frac{300 \ mi}{270 \ mi}$$
$$= 1.1 \ actual \ mi$$
$$\qquad per \ odom \ mi$$

(c) It takes 4 measures of ground coffee to make 7 cups of drink. How many measures should be used for 42 cups?

$$\frac{V_{grind}}{V_{drink}} = \frac{4 \ measures}{7 \ cups}$$

$$V_{grind} = \frac{4 \ meas}{7 \ cups} \ (42 \ cups)$$
$$= 24 \ measures$$

7. When a 200 lb man gets into his car its suspension compresses 1½ inches. When his wife gets in with him the suspension compresses another ¾ inch. How much does his wife weigh? Consider the suspension to be a linear spring.

8. The volume of liquid contained in an oil drum is directly proportional to the depth of the liquid. If there are 23 gallons when the depth is 3 feet, how much liquid is left in the drum when the depth is 6 inches?

9. An important law governing the behavior of gases is Charles' Law which can be stated: "If the pressure of a sample of gas is kept constant, its volume is directly proportional to its absolute temperature." (Absolute temperature T is measured in Kelvin, K.) Apply this law to the following problem.

The temperature of a liter of nitrogen gas boiling off the surface of an open container of liquid nitrogen is 77 K. What is the volume of this gas when it warms up to room temperature (300 K)? (As the gas is not confined, its pressure does not change during warming.)

10. George and Mary fill the gas tank in their car before driving home for school vacation. Every once in a while they fill the tank again as follows:

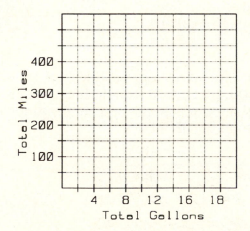

After 110 miles the tank takes 4.3 gallons;
"   250 miles   "   4.0 gallons;
"   400 miles   "   6.1 gallons.

(a) On the graph on the left plot points representing *total* miles driven vs. *total* gas consumed. (b) Draw a straight line through the data points and find its slope. (c) On the basis of the graph determine the average fuel mileage in mi/gal.

11. The mass m of a body is directly proportional to its weight W (see Essay following this Round). Suppose Mr. Smith's mass is 60 kg. A scale shows his weight to be 589 N. After he takes a strenuous hike the scale shows his weight to be 570 N. Use a ratio to determine the mass of slimmed down Mr. Smith.

7. When a 200 lb man gets into his car its suspension compresses 1½ inches. When his wife gets in with him the suspension compresses another ¾ inch. How much does his wife weigh? Consider the suspension to be a linear spring.

$$\frac{Total\ wt}{200\ lb} = \frac{1.5\ in + 0.75\ in}{1.5\ in}$$
$$= 1.5$$
$$Total\ wt = 1.5\ (200\ lb)$$
$$= 300\ lb$$

Thus wife's wt = 300 lb − 200 lb = 100 lb

8. The volume of liquid contained in an oil drum is directly proportional to the depth of the liquid. If there are 23 gallons when the depth is 3 feet, how much liquid is left in the drum when the depth is 6 inches?

$$\frac{V'}{V} = \frac{D'}{D}$$

using D' = 6 in = 0.5 ft we have
$$V' = V\left(\frac{D'}{D}\right) = 23\ gal\left(\frac{0.5\ ft}{3\ ft}\right) = 3.8\ gal$$

9. An important law governing the behavior of gases is Charles' Law which can be stated: "If the pressure of a sample of gas is kept constant, its volume is directly proportional to its absolute temperature." (Absolute temperature T is measured in Kelvin, K.) Apply this law to the following problem.

The temperature of a liter of nitrogen gas boiling off the surface of an open container of liquid nitrogen is 77 K. What is the volume of this gas when it warms up to room temperature (300 K)? (As the gas is not confined, its pressure does not change during warming.)

using primes to refer to warmer gas:
$$V'/V = T'/T$$
$$V' = V(T'/T)$$
$$= 1.0\ \ell\ (300 K / 77 K)$$
$$= 3.9\ \ell$$

10. George and Mary fill the gas tank in their car before driving home for school vacation. Every once in a while they fill the tank again as follows:

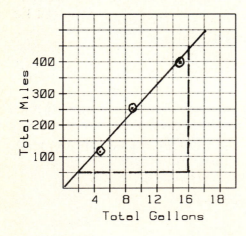

After 110 miles the tank takes 4.3 gallons;
"   250 miles   "   4.0 gallons;
"   400 miles   "   6.1 gallons.

(a) On the graph on the left plot points representing *total* miles driven vs. *total* gas consumed. (b) Draw a straight line through the data points and find its slope. (c) On the basis of the graph determine the average fuel mileage in mi/gal.

(b) Slope = $\frac{\Delta\ miles}{\Delta\ gallons} = \frac{400\ mi}{14\ gal}$
$$= 29\ mi/gal$$

(C) Mileage = slope = 29 mi/gal

11. The mass m of a body is directly proportional to its weight W (see Essay following this Round). Suppose Mr. Smith's mass is 60 kg. A scale shows his weight to be 589 N. After he takes a strenuous hike the scale shows his weight to be 570 N. Use a ratio to determine the mass of slimmed down Mr. Smith.

Primes refer to new mass and weight.
$$\frac{M'}{M} = \frac{W'}{W}$$

$$M = (60\ kg)\ \frac{570\ N}{589\ N} = 58\ kg$$

Smith lost 2 kg.

Review 5 — Linear Equations

THE GENERAL LINEAR EXPRESSION

Form of the equation. The equation describing the direct proportion is just a special case of the linear equation, one of the most useful algebraic relationships for describing physical phenomena. A general form of the linear equation is often written

$$y = mx + b$$

in which m and b are constants which do not depend on the variable quantities y and x. Whenever b is zero, this expression reduces to the equation of direct proportion, with proportionality constant m.

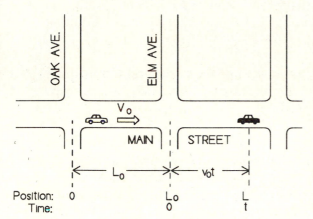

An application from physics. The motion of an object with constant speed along a straight line is easily described using a linear equation.

Imagine a car moving along Main Street with constant speed v_0. Let us use L to denote locations of the car as measured from the intersection of Main Street and Oak Avenue (where L=0). As the car passes the intersection with Elm Avenue (location L_0) a stopwatch to measure time t is turned on. All subsequent locations are given by simply adding to the location at t=0 (which is L_0) additional distances $v_0 t$. Written as an equation we have

$$L = v_0 t + L_0 .$$

In this equation L and t are the dependent and independent variables, respectively. v_0 corresponds to the constant m in the general expression given in the preceding paragraph; L_0 corresponds to b. Negative as well as positive values of t can be used in this equation; when negative values are inserted the calculated values of L are simply the distances of the car from Oak Avenue as it approached the intersection before the stopwatch was turned on.

GRAPHICAL INTERPRETATION

Graph of the general linear equation. In the x-y plane the graph of the linear function y = mx + b is a straight line, but unlike the direct proportion the line does not necessarily pass through the origin. This is illustrated in the figure at the right.

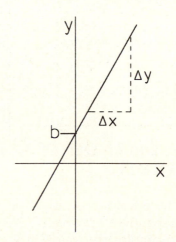

Interpreting the constants. As in the case of direct proportion changes in the independent variable (Δx) are proportional to the corresponding changes in the dependent variable (Δy). Instead of passing through the origin the line intersects the vertical axis at a value y = b, called the "y-intercept." The slope of line (the "rise" Δy divided by the "run" Δx) has the value m:

$$m = \text{slope} = \Delta y / \Delta x .$$

It is important to realize that both constants in the linear equation carry units; b has the same units as y, while the units of m must be a ratio of the units of y and x. In summary:

y = mx + b plots as a straight line; m is the slope of the line and b is the y-intercept.

The following example applies these ideas to the case of the moving car described above:

(See the figure on the previous page.) Suppose that the intersection of Main Street with Elm Avenue is located 50 meters from the intersection with Oak Avenue. The car is travelling at a steady speed of 5.0 m/s.

(a) Write down the linear equation which describes the motion of the car (L vs. t) and draw the corresponding graph. (b) How far is the car from Oak Avenue when the stopwatch reads 30 s?

DISCUSSION: (a) The appropriate equation is $L = v_o t + L_o$

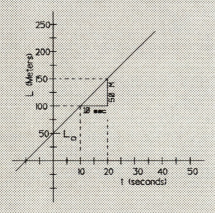

where $v_o = 5.0$ m/s and the position at t = 0 (Elm Ave.), $L_o = 50$ m. For the graph identify the slope with vo; i.e., the rise must have 5.0 units for each 1.0 units of run. For convenience of scale the rise and run are plotted as 50 and 10 respectively, as shown at the left. Also the line must pass through the vertical axis at the intercept $L_o = 50$ m.

(b) For t = 30 s we have $L = v_o t + L_o$
$$= (5.0 \text{ m/s})t + 50.\text{m}$$
$$= (5.0 \text{ m/s})(30 \text{ s}) + 50.\text{m}$$
$$= 200 \text{ meters.}$$

LINEAR DEPENDENCE WHERE THE VARIABLES CHANGE IN OPPOSITE DIRECTIONS

In all of the examples of direct proportion or of the general linear dependence given so far, increases in the independent variable (x) are accompanied by increases in the dependent variable (y). (Likewise in those examples decreases in x correspond to decreases in y.) However, many situations occur in which *decreases* in one variable are proportional to *increases* in the other variable. For such cases the general form of the equations are just as given before:

$$y = kx \quad \text{(direct proportion)}$$
$$y = mx + b \quad \text{(general linear dependence).}$$

However,

> • *when the variables change linearly in opposite directions the proportionality parameters k (or m) are* <u>*negative*</u> *rather than positive numbers.*

Negative slope. For negative k or m linear dependence is also represented by a straight line graph, but the straight line is tilted downwards along the horizontal axis rather than upwards. This is shown in the figure at the right. A positive (upward) "rise" of the line Δy corresponds to a negative (leftward) "run" Δx.

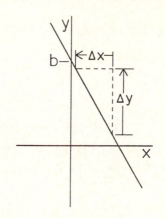

> *Linear (or proportional) dependence in opposite directions corresponds to a straight line graph with a negative slope.*

An example of constant speed motion can be used to illustrate linear but oppositely changing variables. This is the case in which the object moves towards lower values of the position variable as time goes on, i.e., when its motion is described by a negative speed, as in this situation:

Referring to the previous example, the car is returning at the same speed down Main Street moving towards Oak Avenue (v_o = -5.0 m/s). Again we time the motion so t = 0 when the car passes Elm Avenue (L_o = 50 m). When does the car reach Oak Avenue?

DISCUSSION: The question can be restated this way: find t when L = 0. Use the linear equation for the motion as follows:

$$L = v_o t + L_o$$

$$0 = (-5.0 \text{ m/s})t + 50 \text{ m}.$$

The value of t which satisfies this equation is t = 10 s.

Do not confuse this case with that of "inverse proportion," which is an entirely different relationship. Inverse proportion, described by $y = k/x$, will be taken up in Review 14.

Skill Drill 5

This Drill extends your work with direct proportion to embrace general linear phenomena, including practice interpreting graphs. As in the last Drill the purpose of the first question is to remind you of key points covered in the preceding Review.

1. Review of important points:

(a) Write an equation containing two constants expressing a general linear relationship between y and x.

(b) Sketch a graph of this equation, showing the y-intercept, a "rise" Δy, and the corresponding "run" Δx. Assume the constants are both positive. (A freehand sketch will do.)

(c) How are the constants in the equation related to the intercept, Δy, and Δx?

(d) Sketch a graph of a linear equation with both a negative slope and a negative y-intercept.

2. On the graphs provided below draw in lines representing each of the following equations:

(a) y = x + 5   (b) y = −x + 3   (c) y = −2x − 3.

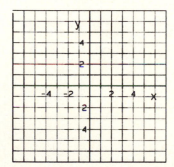 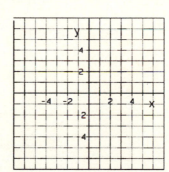 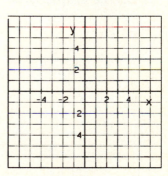

3. The height h above the floor of the platform of a bathroom scale when persons of different weight W step on it is given in this table:

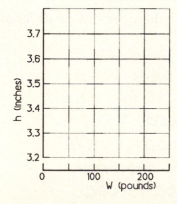

h (inches)	W (pounds)
3.5	100
3.4	150
3.3	200
3.2	250

(a) Plot this data on the graph of h versus W at the left.
(b) Find the slope of the line including correct units and sign.
(c) What is the height of the platform above the floor when nobody is standing on it?

53

Skill Drill 5 — SOLUTIONS AND ANSWERS

1. Review of important points:

(a) Write an equation containing two constants expressing a general linear relationship between y and x.

(b) Sketch a graph of this equation, showing the y-intercept, a "rise" Δy, and the corresponding "run" Δx. Assume the constants are both positive. (A freehand sketch will do.)

(c) How are the constants in the equation related to the intercept, Δy, and Δx?

(d) Sketch a graph of a linear equation with both a negative slope and a negative y-intercept.

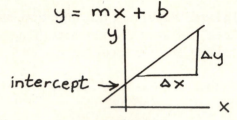

$$y = mx + b$$

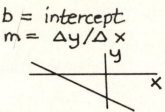

$$b = intercept$$
$$m = \Delta y / \Delta x$$

2. On the graphs provided below draw in lines representing each of the following equations:

(a) y = x + 5   (b) y = −x + 3   (c) y = −2x − 3.

Slope = 1, b = 5   *slope = −1, b = 3*   *slope = −2, b = −3*

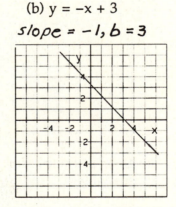
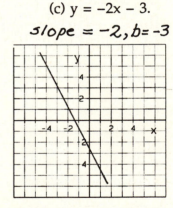

3. The height h above the floor of the platform of a bathroom scale when persons of different weight W step on it is given in this table:

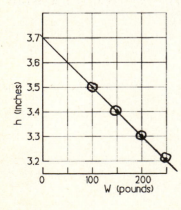

h (inches)	W (pounds)
3.5	100
3.4	150
3.3	200
3.2	250

(a) Plot this data on the graph of h versus W at the left.
(b) Find the slope of the line including correct units and sign.
(c) What is the height of the platform above the floor when nobody is standing on it?

(b) slope, m
$$= \frac{0.30 \ in}{-150 \ lb}$$
$$= -0.0020 \ in/lb$$

(c) The required height is the intercept
$$= 3.7 \ in$$

4. The graph below gives the speed v of a boulder dropped from a castle wall onto troops at the base of the cliff on which the castle stands. The time t refers to the number of seconds after the boulder has passed by the base of the castle.

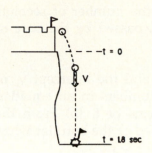

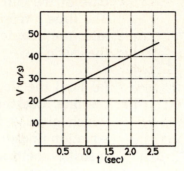

(a) Find the slope g and the intercept v_0 of the line. Use these numbers in an equation which gives v in terms of t. (b) From the equation find the speed of the boulder when it strikes the troops at t = 1.8 s.

5. The depth of distilled water in a laboratory storage jug varies from day to day as follows:

Day	Depth (cm)
2	26
4	22
6	18
10	10

(a) Plot these data on the graph at the right, and draw a straight line through the points. (b) Using the slope and the intercept of your graph write a linear equation representing these data. (c) What might you suppose the depth of water was on day zero?

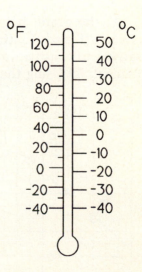

6. Wall thermometers frequently have scales for degrees Celsius and Fahrenheit printed side by side, as illustrated here.

(a) On the graph below draw a straight line representing the relationship between °F and °C. (Label the graph and pick out enough points from the drawing to construct a reasonably accurate line.) (b) Determine the slope and intercept. (c) Using these results write down a formula relating °F and °C.

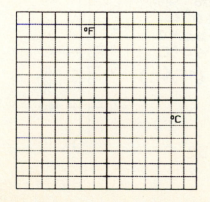

4. The graph below gives the speed v of a boulder dropped from a castle wall onto troops at the base of the cliff on which the castle stands. The time t refers to the number of seconds after the boulder has passed by the base of the castle.

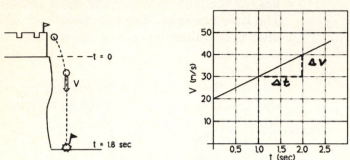

(a) Find the slope g and the intercept v_o of the line. Use these numbers in an equation which gives v in terms of t. (b) From the equation find the speed of the boulder when it strikes the troops at t = 1.8 s.

(a) slope, $g = \dfrac{\Delta V}{\Delta t} = \dfrac{10\,m/s}{1.0\,s}$

$= 10\,m/s^2$

Intercept, $V_o = 20\,m/s$

Thus: $V = (10\,m/s^2)\,t + 20\,m/s$

(b) $V = (10\,m/s^2)(1.8\,s) + 20\,m/s$

$= 38\,m/s$

5. The depth of distilled water in a laboratory storage jug varies from day to day as follows:

Day	Depth (cm)
2	26
4	22
6	18
10	10

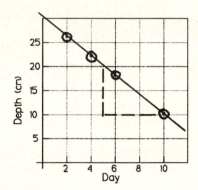

(a) Plot these data on the graph at the right, and draw a straight line through the points. (b) Using the slope and the intercept of your graph write a linear equation representing these data. (c) What might you suppose the depth of water was on day zero?

(b) slope $= \dfrac{-10\,cm}{5.0\,days} = -2.0\,cm/day$; intercept $= 30\,cm$

$D = -(2.0\,cm/day)t + 30\,cm$

(c) 30 cm (intercept)

6. Wall thermometers frequently have scales for degrees Celsius and Fahrenheit printed side by side, as illustrated here.

(a) On the graph below draw a straight line representing the relationship between °F and °C. (Label the graph and pick out enough points from the drawing to construct a reasonably accurate line.) (b) Determine the slope and intercept. (c) Using these results write down a formula relating °F and °C.

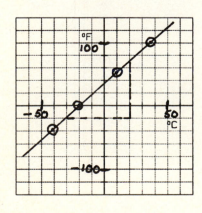

(a) Points: $100°F \rightarrow 38°C$

$50°F \rightarrow 10°C$

$0°F \rightarrow -22°C$

$-40°F \rightarrow -40°C$

(b) slope $= \dfrac{90°F}{50°C} = \dfrac{9}{5}\,(°F/°C)$

intercept $= 32°F$

(c) $T_F = (\frac{9}{5}\,°F/C°)\,T_C + 32°F$

Review 6 — Rearranging Equations: Basic Algebra

To find answers to problems in physics it is often necessary to do some algebraic manipulation. Most of the time the goal is to solve for an unknown, i.e., to isolate on one side of an equation the quantity which is to be determined, leaving the remaining quantities on the other side. Sometimes the purpose is to find an equivalent expression which is in a simpler or more familiar form. In this section the rules for rearranging algebraic expressions are reviewed and illustrated by applying them to linear equations.

MOVING QUANTITIES ACROSS THE EQUAL SIGN

Basic rule of equations. When an equation is rearranged, it is essential that the resulting expression yield the same solution. In other words the transformed equation must be "equivalent" to the original one. This will occur if the following rule is observed.

A mathematical operation which is applied to one side of the equal sign must also be applied to the other side.

Thus multiplying the entire left-hand side ("lhs") by some number or expression is allowed as long as the same multiplication is carried out on the right-hand side ("rhs")— likewise for the operations of division, addition, subtraction, raising to powers, or taking roots. This idea underlies the methods for moving quantities from one side of an equation to the other, outlined below.

Moving factors—cross multiplication. Sometimes we have a factor which multiplies or divides one *entire* side of an equation (not just a term on that side), and we want it to appear instead on the other side. To do this we can multiply or divide both sides by that factor in such a way that it cancels out on the original side. This operation was applied in Review 4 as a way of rearranging ratio expressions. As pointed out in that Review, it is convenient to regard this procedure as a "cross multiplication" which may be summarized this way:

Factors of one side of an equation can be moved diagonally across the equal sign, i.e., from the denominator on one side to the numerator on the other side, or vice versa.

This is illustrated by the example below, where the factor (b+c) is changed from the denominator on the lhs to the numerator on the rhs:

$$\frac{a}{b+c} = \frac{d}{e} \quad \rightarrow \quad a = (b+c)\frac{d}{e}$$

As another example we have

$$\frac{(a)(b)}{c} = \frac{d}{e} \quad \rightarrow \quad a = \frac{(c)(d)}{(b)(e)}$$

Moving terms—transposition. We can also move *terms* from one side of an equation to the other. Suppose, for example, in the equation

$$a + b = c$$

we wish to move the term b to the rhs, leaving term a isolated on the lhs. This is done by subtracting b from both sides to get

$$a + b - b = c - b \quad or \quad a = c - b.$$

This way of rearranging the equation can be conveniently thought of as a transposition of b from one side to the other with a change of sign, like this:

$$a + b = c - \qquad or \qquad a = c - b.$$

The rule for transposition is simply stated:

Terms can be moved from one side of an equation to the other if accompanied by a change of sign.

ADDING, SUBTRACTING, MULTIPLYING, AND DIVIDING

Combining signs. Factors can be positive or negative. The following rule applies:

Multiplying or dividing numbers of like sign yields a positive result; otherwise the result is negative.

Thus, for example, the following three products all have the same value (+ab):

$$(+a)(+b) \;=\; (-a)(-b) \;=\; -(+a)(-b) \,.$$

Groups of quantities—associative rule. This paragraph and the following ones have to do with performing basic algebraic manipulations when there are groups of quantities in an expression, either a series of terms or strings of factors. The first rule has to do with how we can rearrange the parentheses which group the quantities together.

In a sum of three or more terms or a product of three or more factors it doesn't matter how we group the quantities within parentheses.

Subtraction and division are also covered by this rule since subtraction is simply addition of a negative number, and division is simply multiplication by the reciprocal (reciprocal $x \equiv 1/x$). The rule applied to addition and subtraction is illustrated by the following examples involving quantities a, b, and c:

$$a + b + c = (a + b) + c = a + (b + c)$$
$$a + b - c = a + (b - c).$$

A note of caution: In grouping terms within parentheses be careful about which terms are actually being subtracted. For instance a-b+c equals a+(-b+c); it does *not* equal a-(b+c).

The associative rule applied to multiplication and division is illustrated by these examples:

$$abc = (ab)c = a(bc)$$
$$(ab)/c = a(b/c) \,.$$

Groups of quantities—commutative rule.

> *Addition (and subtraction) or multiplication of a series of numbers doesn't depend on the order in which the operations take place.*

This rule is also illustrated by combinations of a, b, and c:

$$a + b - c = a - c + b = b - c + a$$

and

$$abc = acb = bca \,.$$

Groups of quantities—distributive rule. This has to do with the way in which a complicated expression involving sums of terms inside parentheses (polynomials) can be reduced to a sum of relatively simple terms.

> *Multiplying (or dividing) a quantity consisting of a sum of terms inside a parenthesis by a number is the same as multiplying (or dividing) each of the terms by that number.*

This is illustrated using quantities a, b, c, d as follows:

$$(a + b - c)d = ad + bd - cd$$
$$(a - b + c)/d = a/d - b/d + c/d \,.$$

Numerical examples. The above rules for manipulating groups of numbers may seem arbitrary and abstract, but actually they are simply statements about basic facts of arithmetic and the conventions for using parentheses. Any of these rules can be easily verified using numerical values, as in the following computations:

<u>Association</u>

$$2+7+10 = (2+7)+10 = 19 = 2+17 = 2+(7+10)$$
$$(3\times8)/6 = 24/6 = 4 = 3(4/3) = 3(8/6)$$

<u>Commutation</u>

$$12+7-5 = 19-5 = 14 = 7+7 = 12-5+7$$
$$8\times10\times2 = 80\times2 = 160 = 20\times8 = 10\times2\times8$$

<u>Distribution</u>

$$(20+1)2 = 21\times2 = 42 = 40+2 = 20\times2 + 1\times2$$
$$(11-4)/7 = 7/7 = 1 = 7/7 = 11/7 - 4/7$$

Multiplying polynomials by one another. A common algebraic situation is when two (or more) sums of terms within parentheses (polynomials) are multiplied together. The above principles of association, commutation, and distribution can be invoked (see exercise in Skill Drill 6) to show that this multiplication can be summarized in the following relatively simple rule:

> *Multiply each term of one polynomial by each term of the other polynomial, and add the products together.*

Thus for example:

$$(a + b + c)(d + e) = ad + bd + cd + ae + be + ce .$$

Some particularly important special cases whose results are well worth remembering are the following:

$$(a \pm b)^2 = a^2 + b^2 \pm 2ab \quad and \quad (a + b)(a - b) = a^2 - b^2 .$$

STEPS IN SOLVING FOR AN UNKNOWN—AN EXAMPLE DESCRIBING MOTION

In rearranging an equation to solve for an unknown, two things occur:

(1) *All terms or factors containing the unknown are moved to the lhs, then*
(2) *all other terms or factors either cancel or are moved to the rhs.*

The following problem shows how these two steps are typically carried out using the rules of algebraic manipulation which have been summarized in this Review.

The average speed v_{av} (v for "velocity") of a car moving from location L_1 to location L_2 is found by dividing the distance between the two positions by the time interval $(t_2 - t_1)$ elapsed during the motion. Expressed as an equation

$$v_{av} = \frac{L_2 - L_1}{t_2 - t_1}$$

Suppose the car is at $L_1 = 100$ m when a stopwatch reads 10 s. If the average speed is 10 m/s, what does the stopwatch read when the car is at $L_2 = 1000$ m ? In other words, what is t_2?

DISCUSSION: We are asked to solve for t_2, which requires that by algebraic manipulation we isolate it on the left hand side (lhs) of the expression.

The first step is to bring t_2 into the numerator on the lhs. This is done by *cross multiplying* with the factor (t_2-t_1). Then the *distributive rule* allows us to write

$$t_2 v_{av} - t_1 v_{av} = L_2 - L_1 .$$

After this all quantities (besides t_2) must be moved to the right hand side (rhs). Begin by *transposing* $(-t_1 v_{av})$ to the rhs yielding

$$t_2 v_{av} = L_2 - L_1 + t_1 v_{av}.$$

Next *cross multiply* to bring the v_{av} on the lhs to the denominator of the rhs. Finally dividing by that v_{av} according to the *distributive rule* gives us

$$t_2 = \frac{L_2 - L_1}{v_{av}} + t_1.$$

The final quantitative answer is obtained by substituting the given numerical information: the car passes the 1000 m location when the stopwatch reads

$$t_2 = (1000 \text{ m} - 100 \text{ m})/(10 \text{ m/s}) + 10 \text{ s} = 100 \text{ seconds}.$$

Skill Drill 6

The following exercises should help bring you up to speed doing basic algebraic manipulations. Questions in the first part of the Drill involve only abstract symbols and numbers, typical of problems you find in mathematics courses. But in the spirit of problem solving as you will find it in physics there are also a number of word problems and quantities with units.

1. This exercise is a tour of the main points covered in Review 6.

(a) Using cross multiplication find factor e in terms of the other quantities in the expression: $(a+b)/c = d/e$.

(b) Transpose terms so that d alone appears on the lhs of the equation: $a+b-c = -d+e$.

(c) Rearrange the parentheses in each of the following expressions so as to put it in the form indicated to its right. In each case what is X in terms of a, b, c, and d?

$$(a-b-c/2)+d \rightarrow a-b+(X)$$
and
$$(abc)/d \rightarrow ab/(X) .$$

(d) Using numerical values a=3, b=4, c=-5, and d=5 verify that

$$\frac{(a+b+c)d}{abc} = \frac{d}{bc} + \frac{d}{ac} + \frac{d}{ab}$$

(e) Using numerical values a=3, b=4, c=-5, and d=5 verify that

$$(a+b)(c+d) = ac+bc+ad+bd .$$

(f) Using the distributive rule show that

$$(a+b)(c+d) = ac + bc + ad + bd$$

(g) Using the result of part (f) obtain the following useful "binomial expansions:"

$$(x-y)^2 = x^2 + y^2 - 2xy$$
and
$$(x+y)(x-y) = x^2 - y^2 .$$

2. Find a factor which is common to every term in each of the following expressions. Rewrite each expression with the factor separated from other terms.

(a) $4ab + 2bc$

(b) $2b^2 - 6ab$

Skill Drill 6 — SOLUTIONS AND ANSWERS

1. This exercise is a tour of the main points covered in Review 6.

(a) Using cross multiplication find factor e in terms of the other quantities in the expression: $(a+b)/c = d/e$.

$$e = \frac{c}{(a+b)}$$

(b) Transpose terms so that d alone appears on the lhs of the equation: $a+b-c = -d+e$.

$$d = e - a - b + c$$

(c) Rearrange the parentheses in each of the following expressions so as to put it in the form indicated to its right. In each case what is X in terms of a, b, c, and d?

$$(a-b-c/2)+d \rightarrow a-b+(X)$$

and

$$(abc)/d \rightarrow ab/(X) .$$

$$a - b + \left(-c/2 + d\right)$$
$$\rightarrow X = d - c/2$$
and
$$ab(c/d) = ab/\left(\frac{d}{c}\right)$$
$$\rightarrow X = d/c$$

(d) Using numerical values a=3, b=4, c=-5, and d=5 verify that

$$\frac{(a+b+c)d}{abc} = \frac{d}{bc} + \frac{d}{ac} + \frac{d}{ab}$$

$$\frac{(3+4+5)5}{3\cdot4\cdot5} = \frac{12\cdot5}{60} = 1$$

also

$$\frac{5}{4\cdot5} + \frac{5}{3\cdot5} + \frac{5}{3\cdot5} = \frac{1}{4} + \frac{1}{3} + \frac{5}{12}$$
$$= (3+4+5)/12 = 1$$

(e) Using numerical values a=3, b=4, c=-5, and d=5 verify that

$$(a+b)(c+d) = ac+bc+ad+bd .$$

also
$$(3+4)(5+5) = 7(10) = 70$$
$$3\cdot5 + 4\cdot5 + 3\cdot5$$
$$= 15 + 20 + 15 = 70$$

(f) Using the distributive rule show that

$$(a+b)(c+d) = ac + bc + ad + bd$$

$$(a+b)c + (a+b)d$$
$$= ac + bc + ad + bd$$

(g) Using the result of part (f) obtain the following useful "binomial expansions:"

$$(x-y)^2 = x^2 + y^2 - 2xy$$

and

$$(x+y)(x-y) = x^2 - y^2 .$$

substituting: $a=x, b=-y, c=x, d=-y$
$$(x-y)(x-y) = x^2 - yx - xy + y^2$$
substituting: $a=x, b=y, c=x, d=-y$
$$(x+y)(x-y) = x^2 + xy - xy - y^2$$

2. Find a factor which is common to every term in each of the following expressions. Rewrite each expression with the factor separated from other terms.

(a) $4ab + 2bc$

common factor: $2b \rightarrow 2b(2a+c)$

(b) $2b^2 - 6ab$

common factor: $2b \rightarrow 2b(b-3a)$

(c) $2x^2y + 4bxy + 2y^2x$

(d) $3x(a^2 + 2ay + y^2) + 6x^2(y + a)$
   (NOTE: In this expression the common factor contains
   a binomial.)

3. Expand the following expressions into a polynomial in x by carrying out the indicated multiplications and grouping together like powers of x.

(a) $x(4x+3ax+2)$

(b) $(x+2)(x-5)$

(c) $x(x^2-3x+6)$

(d) $(2-x)(2x^2+4x-2)$

4. Solve the following equations. (Isolate x on the lhs, then calculate its numerical value.)

(a) $4x + 12 = x + 30$

(b) $4x - 3 = 5(x+3)$

(c) $\dfrac{5(x+4)}{3} = 8 + \dfrac{3x}{2}$

(d) $0.75x - 0.25x - 2.00 = 0.25x + 3.00$

5. Set up an equation to represent each of the following problems and solve for the unknown. Be sure to include units in the solution.

(a) Admission to the college play costs $5.00 for the general public and $3.00 for students and faculty. Half as many tickets were sold to students and faculty as to the general public, and a total of $455 was collected. How many general admission tickets were sold?

(b) You buy two certificates of deposit, one of which yields 9.0% interest annually, while the other earns 8.5%. The amount invested in the second C.D. is twice the amount invested in the first. If the total amount earned in one year is $780, how much is the value of the first C.D.?

(c) $2x^2y + 4bxy + 2y^2x$ *Common factor: $2xy \longrightarrow 2xy(x+2b+y)$*

(d) $3x(a^2 + 2ay + y^2) + 6x^2(y + a)$

    (NOTE: In this expression the common factor contains a binomial.)

Rewrite: $3x(a+y)^2 + 6x^2(y+a)$
$$= 3x(a+y)[(a+y)+2x]$$
$$\longrightarrow \text{Common factor} = 3x(a+y)$$

3. Expand the following expressions into a polynomial in x by carrying out the indicated multiplications and grouping together like powers of x.

(a) $x(4x+3ax+2)$
$$4x^2 + 3ax^2 + 2x = (4+3a)x^2 + 2x$$

(b) $(x+2)(x-5)$
$$x^2 + 2x - 5x - 10 = x^2 - 3x - 10$$

(c) $x(x^2-3x+6)$
$$x^3 - 3x^2 + 6x$$

(d) $(2-x)(2x^2+4x-2)$
$$4x^2 + 8x - 4 - 2x^3 - 4x^2 + 2x$$
$$= -2x^3 + 12x - 4$$

4. Solve the following equations. (Isolate x on the lhs, then calculate its numerical value.)

(a) $4x + 12 = x + 30$
$$3x = 18 \longrightarrow x = 6$$

(b) $4x - 3 = 5(x+3)$
$$4x - 5x = 15 + 3 \longrightarrow -x = 18$$
$$x = -18$$

(c) $\dfrac{5(x+4)}{3} = 8 + \dfrac{3x}{2}$
$$\frac{5}{3}x - \frac{3}{2}x = -\frac{20}{3} + 8$$
$$\frac{10-9}{6}x = -\frac{20+24}{3}$$
$$x/6 = 4/3 \longrightarrow x = 8$$

(d) $0.75x - 0.25x - 2.00 = 0.25x + 3.00$
$$0.50x - 0.25x = 3.00 + 2.00$$
$$0.25x = 5.00 \longrightarrow x = 20.$$

5. Set up an equation to represent each of the following problems and solve for the unknown. Be sure to include units in the solution.

(a) Admission to the college play costs $5.00 for the general public and $3.00 for students and faculty. Half as many tickets were sold to students and faculty as to the general public, and a total of $455 was collected. How many general admission tickets were sold?

G = no of tickets to general public
$$(\$5.00/ticket)\,G + (\$3.00/ticket)\,G/2 = \$455$$
$$(\$6.50/ticket)\,G = \$455$$
$$G = 70 \text{ tickets}$$

(b) You buy two certificates of deposit, one of which yields 9.0% interest annually, while the other earns 8.5%. The amount invested in the second C.D. is twice the amount invested in the first. If the total amount earned in one year is $780, how much is the value of the first C.D.?

V = value of 1st CD
$$(0.090)\,V + (0.085)(2v) = \$780 \longrightarrow 0.26v = \$780$$
$$\text{Thus } V = \$3000$$

(c) Two planes, one in New York and one in London, take off at the same time heading for each other's city, 3469 miles apart. The eastbound plane, travelling with the jet stream, flies at 700 mph while the westbound plane travels at 600 mph. How long after taking off do they meet?

6. The following problems have a scientific context. You do not need to understand all the details, but you should be able to *rearrange each of the given equations algebraically* as directed.

(a) The location of the "center of mass" X_{cm} of a molecule consisting of two atoms of masses m_1 and m_2 is given by

$$(m_1+m_2)X_{cm} = m_1x_1 + m_2x_2$$

where X_{cm} and the position of the atoms x_1 and x_2 are measured from some common origin. Find m_1 in terms of the other quantities.

(b) When a marble of mass m having a speed v_1 rolls up an incline its speed decreases to v_2. If the height of the incline is h the two speeds are related very closely by the equation

$$\tfrac{1}{2}mv_1^2 - \tfrac{1}{2}mv_2^2 = mgh$$

where g is a constant (gravitational acceleration). Write an expression for v_1 in terms of the other quantities.

(c) The ability of an electrical component to permit a flow of electricity through it is described by a quantity called resistance R.
When three components are connected in a certain standard way called a "parallel connection" the ability of the entire arrangement to conduct electricity R_{eq} is related to the individual resistances R_1, R_2, and R_3 by the equation

$$\frac{1}{R_{eq}} = \frac{1}{R_1} + \frac{1}{R_2} + \frac{1}{R_3} .$$

(R_{eq} means "equivalent resistance.") Show that this expression can be rewritten as

$$R_{eq} = \frac{R_1R_2R_3}{R_2R_3 + R_3R_1 + R_1R_2} .$$

(c) Two planes, one in New York and one in London, take off at the same time heading for each other's city, 3469 miles apart. The eastbound plane, travelling with the jet stream, flies at 700 mph while the westbound plane travels at 600 mph. How long after taking off do they meet?

$$NY \xrightarrow{\quad 3469\,mi \quad} \cdot L$$
$$\text{(700 mph)}t \quad \text{(600 mph)}t$$

$$(700\,mi/hr)^t + (600\,mi/hr)^t$$
$$= 3469\,mi$$
$$(1300\,mi/hr)^t = 3469\,mi$$
$$\longrightarrow t = 2.67\,hr = 2\,hr\ 40\,min$$

6. The following problems have a scientific context. You do not need to understand all the details, but you should be able to *rearrange each of the given equations algebraically* as directed.

(a) The location of the "center of mass" X_{cm} of a molecule consisting of two atoms of masses m_1 and m_2 is given by

$$(m_1+m_2)X_{cm} = m_1x_1 + m_2x_2$$

where X_{cm} and the position of the atoms x_1 and x_2 are measured from some common origin. Find m_1 in terms of the other quantities.

$$m_1 X_{cm} + m_2 X_{cm} = m_1 x_1 + m_2 x_2$$
$$m_1 X_{cm} - m_1 X_1 = m_2 X_2 - m_2 X_{cm}$$
$$m_1 (X_{cm} - x_1) = m_2 (x_2 - X_{cm})$$

$$m_1 = \frac{x_2 - X_{cm}}{X_{cm} - x_1}$$

(b) When a marble of mass m having a speed v_1 rolls up an incline its speed decreases to v_2. If the height of the incline is h the two speeds are related very closely by the equation

$$\tfrac{1}{2}mv_1^2 - \tfrac{1}{2}mv_2^2 = mgh$$

where g is a constant (gravitational acceleration). Write an expression for v_1 in terms of the other quantities.

$$\tfrac{1}{2}mv_1^2 = mgh + \tfrac{1}{2}mv_2^2$$

$$v_1^2 = \frac{2\cancel{m}gh + \cancel{m}v_2^2}{\cancel{m}}$$

$$v_1 = \sqrt{2gh + v_2^2}$$

(c) The ability of an electrical component to permit a flow of electricity through it is described by a quantity called resistance R.
When three components are connected in a certain standard way called a "parallel connection" the ability of the entire arrangement to conduct electricity R_{eq} is related to the individual resistances R_1, R_2, and R_3 by the equation

$$\frac{1}{R_{eq}} = \frac{1}{R_1} + \frac{1}{R_2} + \frac{1}{R_3}.$$

(R_{eq} means "equivalent resistance.") Show that this expression can be rewritten as

$$R_{eq} = \frac{R_1 R_2 R_3}{R_2 R_3 + R_3 R_1 + R_1 R_2}.$$

$$\frac{1}{R_{eq}} = \frac{R_2 R_3}{R_1 R_2 R_3} + \frac{R_3 R_1}{R_1 R_2 R_3} + \frac{R_1 R_2}{R_1 R_2 R_3}$$
$$\frac{R_1 R_2 R_3}{R_{eq}} = R_2 R_3 + R_3 R_1 + R_1 R_2$$

Thus

$$R_{eq} = \frac{R_1 R_2 R_3}{R_2 R_3 + R_3 R_1 + R_1 R_2}$$

Review 7 — Simultaneous Linear Equations

It would be convenient if every physics problem could be solved by finding an equation in which the unknown quantity was directly related to all the known quantities. Unfortunately in some cases more than one unknown quantity is involved in the algebraic statement of the problem. However, the problem often can still be solved using additional independent equations bearing on the physics. The main requirement is that there be at least as many of these "simultaneous equations" as there are unknown quantities. This Review discusses how this is done for situations in which all the equations are linear.

TWO SIMULTANEOUS LINEAR EQUATIONS

An example using a linear equation describing simple motion. An example of a linear equation given in Review 5 described the position L of an object moving in a straight path with constant speed v_o. It was written

$$L = v_o t + L_o$$

where L_o is the position when the time $t = 0$. Now consider a case in which two such equations are used to describe a situation, as follows:

Bob and Ray are each driving their pickups down the freeway in the same direction at constant speeds. Bob passes milepost 10 going 50 mi/hr at exactly 2:00 PM. At that same moment Ray passes milepost 8 going 60 mi/hr. At what milepost position does Ray overtake Bob? At what time does this occur?

DISCUSSION: We can write one equation by using information about the motion of Bob's car and a second one by using information about Ray's car. Choose 2:00 PM to be time $t = 0$. Using x to label position instead of L (a common convention), the motion of each of the two cars is described by these linear equations:

$$x_B = (50 \text{ mi/hr})t + 10.0 \text{ mi}$$

and

$$x_R = (60 \text{ mi/hr})t + 8.0 \text{ mi} .$$

According to the problem at a certain time t Bob and Ray are at the same place (call this simply x). Thus at that particular time and place the following two equations hold true:

$$x = (50 \text{ mi/hr})t + 10.0 \text{ mi}$$

and

$$x = (60 \text{ mi/hr})t + 8.0 \text{ mi} .$$

This is a pair of simultaneous linear equations in two unknowns (x and t) which can be solved to give the answers to the problem.

Basic strategy for finding the solutions. Described in broad terms, solving a pair of simultaneous linear equations in two unknowns proceeds as follows:

The two equations are reduced to a single equation in one unknown; the solution of this equation is then substituted back into one of the original equations to produce a second equation, which is then solved for the other unknown.

Two essentially equivalent procedures for carrying out this strategy are given below.

Solving by substitution of an equation. A straightforward approach which can be readily applied in most cases has these steps:

> *(1) Rearrange one equation so that one unknown is given in terms of the other (in the example above, for instance, t in terms of x); then*
> *(2) use that expression to completely replace one of the unknowns in the other equation (for example, replace t).*

The resulting equation contains only one unknown. Now carry out the remainder of the strategy which was summarized above, as follows:

> *(3) Solve the equation found in step (2) for the unknown (x).*
> *(4) Substitute the value which was found for this unknown into one of the original equations, and*
> *(5) solve this equation for the other unknown (t).*

Here is how these steps are applied to the example above:

DISCUSSION: Refer to the two simultaneous equations in the previous example. *(Step 1)* Using the second equation to get t in terms of x we have

$$t = \frac{x - 8.0\,mi}{60\,mi/hr}$$

(Step 2) This is substituted into the first equation to get

$$x = \frac{(50\,mi/hr)(x - 8.0\,mi)}{60\,mi/hr} + 10\,mi$$

or

$$x = (5/6)x - (20/3)\,mi + 10\,mi \,.$$

(Step 3) The solution of this equation is x = 20 mi (mile post 20).

(Step 4) To find t substitute this value of x into the first of the simultaneous equations to get

$$20\,mi = (50\,mi/hr)t + 10\,mi$$

which can be solved *(Step 5)* to give

$$t = 1/5\,hr = 12\,minutes.$$

Ray overtakes Bob at 2:12 PM.

Solving by addition or subtraction of equations. An alternative method of solving the pair of simultaneous equations differs from the one outlined above only in its approach to eliminating one of the unknowns (Steps 1 and 2). This technique works nicely when the equations are relatively simple, and usually requires less algebraic manipulation than the substitution approach:

(1) Rearrange the two equations so that the dependence on one of the variables is contained in each of them in a single term of the same magnitude (negative or positive size).

(Usually this step is carried out by multiplying one of the equations by an appropriate factor which makes one of terms equal in size to the corresponding term in the other equation.) Next

(2) add or subtract the equations so as to eliminate the terms of equal magnitude.

The resulting equation contains just one unknown. The remainder of the procedure is just the same as outlined before (Steps 3–5). Here is this method applied to the previous example:

DISCUSSION: Referring to the pair of simultaneous equations in the example on p. 67 containing the two variables x and t, (*Step 1*) multiply the first equation by the fraction 6/5 so that the terms in t in both equations are of equal size. The two equations are now

$$(6/5)x = (60 \text{ mi/hr})t + 12 \text{ mi}$$

and

$$x = (60 \text{ mi/hr})t + 8.0 \text{ mi} .$$

Upon subtracting the second equation from the first (*Step 2*) the terms in t disappear leaving only one equation in x, viz.

$$(1/5)x = 4.0 \text{ mi}.$$

Thus (*Step 3*)

$$x = 20 \text{ miles},$$

as before.

SIMULTANEOUS EQUATIONS REPRESENTED AS GRAPHS

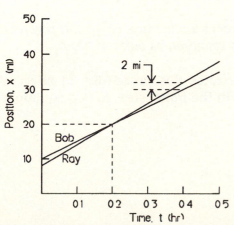

Graphs of equations can provide useful insight into the physics of a problem. A useful representation of a pair of simultaneous equations is a pair of curves plotted on the same graph. If there are just two variables the points where the curves cross each other correspond to the solutions of the equation pair.

Example of two linear equations. The example of the overtaking cars discussed above can be graphed as shown on the left. Each line represents one of the equations in the example. The equations are satisfied simultaneously for the pair of values where the lines cross (x = 20 mi, t = 12 min).

A graph like the one shown above can be used to answer other factual questions. For instance, if we were interested in when Ray will be 2 miles ahead of Bob, we can read this off directly by noting a value of t at which the lines are separated by 2 miles. (The answer is 24 minutes).

Possible solutions. The graphs representing two simultaneous equations also can provide insight into whether or not two solutions are to be expected.

If it turns out that the curves representing a pair of equations in two unknowns do not cross one another there can be no solution at all. For example, in the case of moving cars just discussed, if the two cars travel at the same speed the lines representing their motion would have the same slope; the lines would be parallel and never intersect. One car could never catch up with the other.

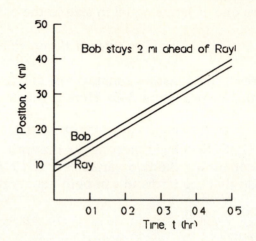

In contrast to the case of pairs of equations with no solutions are situations in which the corresponding curves cross at more than one point (more than one solution pair). This may happen when the equations are more complicated than linear. Nonlinear equations are the topic for a later Review.

Finally, it is worth realizing that if two curves fall directly on top of one another, the two equations they represent are, in fact, equivalent to one another. In such a case the two equations cannot be regarded as "independent" equations; they both contribute identically the same information to the solution of the problem. We would have to look for another relationship among the variables to find a solution.

MORE THAN TWO SIMULTANEOUS LINEAR EQUATIONS

Sometimes in setting up a problem you may need more than two simultaneous equations to represent the given information. In general, as long as there are as many independent equations as there are variables, a solution is possible. The procedure is, in essence, the same as that used to solve a pair of simultaneous equations: the set of equations is reduced to a single equation in one unknown, and the solution of that equation is then substituted back into the original equations. For example, a set of three linear independent equations in three unknowns can be dealt with as follows:

> *Use one of the three equations to eliminate one of the variables from the other two, by substituting for that variable wherever possible; then solve this pair of simultaneous equations as outlined above.*

In fact, the first example problem of this Review actually illustrates this procedure. In effect, two equations containing three variables (x_B, x_R, and t) were reduced, in the initial step, to two equations in two unknowns using a third equation $x_R = x_B = x$.

Skill Drill 7

These problems provide practice in handling simultaneous linear equations, besides giving additional opportunities to work on algebra skills. The first problems are purely algebra exercises covering the techniques in a formal way. However, be sure to do the word problems which follow.

1. Review of key points:

(a) Reduce the linear equation $2x - 3y = -1$ to one unknown by substitution from the equation $x - 2y = -2$.

(b) Referring to the equations in part (a), by what factor can you multiply the second equation in order to eliminate the variable y by adding or subtracting the first equation?

(c) Find x and y.

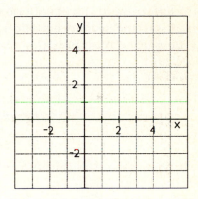

(d) Draw two straight lines representing the pair of equations given in part (a) and verify that the crossing point corresponds to the solution found above.

(e) This question refers to the first example discussed in Review 7. Solve the following set of three simultaneous equations to determine the time t at which Ray is 2.0 miles ahead of Bob:

$$x_B = (50 \text{ mi/hr})t + 10.0 \text{ mi}$$
$$x_R = (60 \text{ mi/hr})t + 8.0 \text{ mi}$$
$$x_R = x_B + 2.0 \text{ mi}.$$

2. Reduce each of the following pairs of equations to a single equation in one variable by substituting one into the other, then solve for x and y.

(a) $3x - 5y = 4$
        $x = -y$

(b) $x + y = 0$
        $5x = 10 - y$

Skill Drill 7 — SOLUTIONS AND ANSWERS

1. Review of key points:

(a) Reduce the linear equation $2x - 3y = -1$ to one unknown by substitution from the equation $x - 2y = -2$.

Write 2nd equation as:
$$X = 2y - 2$$
After substitution the 1st eqn is: $2(2y-2) - 3y = -1$ *or* $4 - y - 3y - 4 = -1$

(b) Referring to the equations in part (a), by what factor can you multiply the second equation in order to eliminate the variable y by adding or subtracting the first equation?

Multiply 2nd eqn by 3/2

(c) Find x and y.

From part (a)
$$y = -1 + 4 \longrightarrow y = 3$$
Inserting this into 2nd eqn
$$X - 6 = -2 \longrightarrow X = 4$$

(d) Draw two straight lines representing the pair of equations given in part (a) and verify that the crossing point corresponds to the solution found above. *To get slopes and intersepts write eqns as* $y = 2x/3 + 1/3$ *and* $y = x/2 + 1$. *Crossing point (4,3) is noted on graph.*

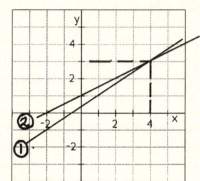

(e) This question refers to the first example discussed in Review 7. Solve the following set of three simultaneous equations to determine the time t at which Ray is 2.0 miles ahead of Bob:

$$x_B = (50 \text{ mi/hr})t + 10.0 \text{ mi}$$
$$x_R = (60 \text{ mi/hr})t + 8.0 \text{ mi}$$
$$x_R = x_B + 2.0 \text{ mi.}$$

Substitute 3rd eqn into 2nd eqn to get:
$$X_B + 2.0 \text{ mi} = (60 \text{ mph}) t + 8.0 \text{ mi}$$
Subtract this from the 1st eqn:
$$-2.0 \text{ mi} = -(10 \text{ mph})t + 2.0 \text{ mi}$$
or
$$t = 0.40 \text{ hr} = 24 \text{ min}$$

2. Reduce each of the following pairs of equations to a single equation in one variable by substituting one into the other, then solve for x and y.

(a) $3x - 5y = 4$
       $x = -y$

Substitute 2nd eqn: $-3y - 5y = 4$
$$-8y = 4 \rightarrow y = -1/2$$
Putting this result into 2nd eqn:
$$\longrightarrow X = 1/2$$

(b) $x + y = 0$
       $5x = 10 - y$

Substitute 1st eqn into 2nd:
$$5x = 10 - (-x)$$
$$4x = 10 \rightarrow X = 2\frac{1}{2}$$
From 2nd eqn: $y = -2\frac{1}{2}$

(c) $y = -\dfrac{1}{3}x + 6$

     $y = -\dfrac{7}{6}x - 4$

3. Reduce each of the following equation pairs to a single equation in one variable by addition or subtraction of equations, then solve for x and y.

(a) $3x + 6y = -15$
     $5x - 3y = 1$

(b) $y = 7x + 3$
     $3y = 20x + 11$

4. On the graph provided draw pairs of lines representing the pairs of equations given in previous exercise (a), and verify your solution.

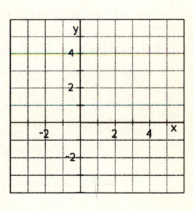

5. Solve each of the following sets of three simultaneous linear equations:

(a) $x + y + z = 4$
     $x - y + z = 0$
     $x - y - z = -2$

(b) $m - n = 2$
     $n + p = -1$
     $m - p = 6$

(c) $y = -\frac{1}{3}x + 6$

Substituting 2nd eqn into 1st eqn:
$$-7/6 x - 4 = -\frac{1}{3}x + 6$$
$$-5/6 x = 10 \rightarrow x = -12$$

$y = -\frac{7}{6}x - 4$

Putting this result into 1st eqn:
$$y = \frac{12}{3} + 6 = 10.$$

3. Reduce each of the following equation pairs to a single equation in one variable by addition or subtraction of equations, then solve for x and y.

Multiplying 2nd eqn by 2 yields pair:

(a) $3x + 6y = -15$
$5x - 3y = 1$

$$3x + 6y = -15$$
$$10x - 6y = 2$$

Adding these:
$$13x = -13 \rightarrow x = -1$$

Using this in 2nd eqn:
$$-5 - 3y = 1 \rightarrow y = -2$$

(b) $y = 7x + 3$
$3y = 20x + 11$

Multiplying 1st eqn by 3:
$$3y = 21x + 9$$
$$3y = 20x + 11$$
Subtracting

$$0 = x - 2 \rightarrow x = 2$$
First eqn:
$$y = 14 + 3 = 17$$

4. On the graph provided draw pairs of lines representing the pairs of equations given in previous exercise (a), and verify your solution.

Rewrite the 2 equations:
$$6y = -3x - 15 \rightarrow y = -\frac{1}{2}x - 5/2$$
$$-3y = -5x + 1 \rightarrow y = \frac{5}{3}x - \frac{1}{3}$$

Slopes and intercepts are used to plot lines as shown $\longrightarrow$

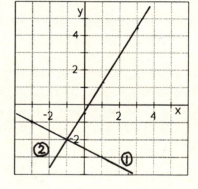

Substitute $x = -1, y = -2$ into each eqn:
$$3(-1) + 6(-2) = -15 \rightarrow -15 = -15$$
$$5(-1) - 3(-2) = 1 \rightarrow 1 = 1$$

5. Solve each of the following sets of three simultaneous linear equations:

(a) $x + y + z = 4$
$x - y + z = 0$
$x - y - z = -2$

Use 2nd eqn $(x = y - z)$ in 1st and 3rd eqns:
$$y - z + y + z = 4 \rightarrow y = 2$$
$$y - z - y - z = -2 \rightarrow z = 1$$
Putting these results into 1st eqn:
$$x = 2 - 1 = 1$$

(b) $m - n = 2$
$n + p = -1$
$m - p = 6$

Substituting from 3rd eqn $(m = p + 6)$:
$$p + 6 - n = 2$$
$$n + p = -1$$
adding:
$$2p + 6 = 1 \rightarrow p = 2\frac{1}{2}$$

$m = -2\frac{1}{2} + 6$
$= 3\frac{1}{2}$
$n = m - 2$
$= 1\frac{1}{2}$

6. Write down two or more independent linear equations for each of the following problems and solve. Indicate units.

(a) You wish to mix A_1 quarts of 1.0% (butterfat content) milk with A_4 quarts of regular 3.8% milk to obtain one gallon of 2.0% milk for breakfast. Find A_1 and A_4. (HINT: The amount of butterfat in the final mixture is the total amount of butterfat in the two portions which are added together.)

(b) A jar of mixed coins contains P pennies, N nickels, and D dimes. The number of pennies equals the number of nickels and dimes put together. There are 50 more pennies than nickels and twice as many nickels as dimes. Find P, N, and D as well as the total value of the coins.

(c) Al makes 5 round trips between the Chicago home office and San Francisco each year, while Mary makes 10 round trips from the home office to Washington, D.C. during the same period. The distance D_W from Chicago to Washington is one third the distance D_S from Chicago to San Francisco. Together Al and Mary cover 30,000 miles on these trips. What are the distances D_S and D_W?

(d) A cruise ship departs Miami harbor with a constant speed of 25 knots (nautical miles per hour). Forty minutes later a Coast Guard cutter from Miami starts after the ship going 45 knots. How much time t after the ship departs does the cutter catch up with it, and how far out to sea D is it at that time?

6. Write down two or more independent linear equations for each of the following problems and solve. Indicate units.

(a) You wish to mix A_1 quarts of 1.0% (butterfat content) milk with A_4 quarts of regular 3.8% milk to obtain one gallon of 2.0% milk for breakfast. Find A_1 and A_4. (HINT: The amount of butterfat in the final mixture is the total amount of butterfat in the two portions which are added together.)

$$\text{Total butterfat} = (0.010)A_1 + (0.038)A_4 = (0.020)(4.0qt) \Big\}$$
$$\text{also:} \quad A_1 + A_4 = 4.0qt$$

Substitute into 1st eqn:

$$(0.010)(4qt - A_4) + (0.038)A_4 = 0.080 qt$$
$$(0.0280)A_4 = (0.080 - 0.040)$$
$$A_4 = 1.4qt \text{ and } A_1 = 4.0qt - 1.4qt = 2.6qt$$

(b) A jar of mixed coins contains P pennies, N nickels, and D dimes. The number of pennies equals the number of nickels and dimes put together. There are 50 more pennies than nickels and twice as many nickels as dimes. Find P, N, and D as well as the total value of the coins.

$$\begin{cases} P = N + D \\ P = N + 50 \text{ coins} \\ N = 2D \end{cases}$$

Substitute from 3rd eqn:
$$P = 2D + D = 3D$$
$$P = 2D + 50 \text{ coins}$$

Subtracting :
$$0 = D - 50 \text{ coins}$$
Thus
$$D = 50 \text{ coins}$$
$$N = 2(50c) = 100 \text{ coins}$$
$$P = N + D = 150 \text{ coins}$$
$$\text{Total value} = (P + 5N + 10D)$$
$$= 1150c = \$11.50$$

(c) Al makes 5 round trips between the Chicago home office and San Francisco each year, while Mary makes 10 round trips from the home office to Washington, D.C. during the same period. The distance D_W from Chicago to Washington is one third the distance D_S from Chicago to San Francisco. Together Al and Mary cover 30,000 miles on these trips. What are the distances D_S and D_W?

$$\begin{cases} 5(2D_S) + 10(2D_W) = 30,000 \text{ mi} \\ D_W = D_S/3 \end{cases}$$

Substituting from 2nd eqn:
$$(5)(2)(3D_W) + (10)(2D_W) = 30,000 \text{ mi}$$

$$50 D_W = 30,000 \text{ mi}$$
$$\to D_W = 600 \text{ mi}$$
and
$$D_S = 3D_W$$
$$= 1800 \text{ mi}$$

(d) A cruise ship departs Miami harbor with a constant speed of 25 knots (nautical miles per hour). Forty minutes later a Coast Guard cutter from Miami starts after the ship going 45 knots. How much time t after the ship departs does the cutter catch up with it, and how far out to sea D is it at that time?

Time zero:

Time = 40 min

$$\begin{cases} D = V_s t \\ D = V_c (t - 40 \text{ min}) \end{cases}$$

$$V_s t = V_c t - V_c (40 \text{ min})$$
or $$t = \frac{V_c (40 \text{ min})}{V_c - V_s} = \frac{25}{45 - 25}(40 \text{ min})$$
$$= 90 \text{ min} = 1\frac{1}{2} \text{ hr}$$
and $$D = (25 \text{ naut mi/hr})(1\frac{1}{2} \text{ hr})$$
$$= 37\frac{1}{2} \text{ nautical miles}$$

Second Round Posttest — Optimum test time: 20 minutes or less

This is a brief test to give you a chance to see how much you have improved your facility with certain skills since beginning the second Round of work. In view of the types of examples given in the Reviews and Drills, more emphasis is given to word problems and applications in this test than in the Pretest.

Space is provided for answers but you will need a sheet of paper for most of the work. Keep track of your time and check your answers against those given at the end of this Round.

STARTING TIME_____   ANSWERS

1. On a map of the United States the distance between Miami and Dallas is 14.30 cm. The mileage scale in the corner of the map shows 5.10 cm to be equivalent to 400 mi. Use a ratio to find the distance in miles between the two cities.

_____

2. Express the following statement as an equation:
   The fractional increase in the length of an iron rod $\Delta L/L$ is directly proportional to the rise in temperature of the rod ΔT. The proportionality constant is the expansion coefficient k (units: degrees^{-1}).

_____

3. What is the slope and y-intercept of the line shown in the graph at the right. Use these data to write an equation for the line.

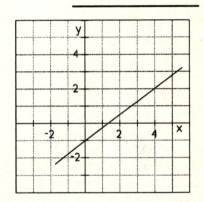

_____

_____

4. Liquid is being poured into a beaker which is sitting on the platform of a laboratory scale. The table below gives the scale readings M (mass in kg) for various volumes V of liquid added.

M (kg)	V (liters)
1.1	1.0
1.5	2.0
1.8	3.0
2.3	4.0
2.6	5.0

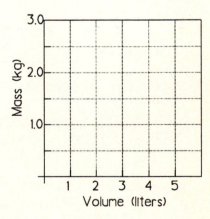

(a) Plot these data on the graph of M versus V and draw a line representing the linear relationship. (b) From the graph determine the constants in the equation $M = mV + M_o$. (c) What are the physical meanings of M_o and m?

_____   _____

77

5. Find the solution of the following equation:

$$\frac{x+2}{4} + \frac{x-2}{2} = 1$$

_____

6. Set up an equation and use it to solve the following problem:

You open both hot and cold water spigots to fill a bath tub. The hot water runs at 2 gal/min, whereas the cold runs at 4 gal/min. How long does it take to run 20 gallons for a bath?

_____   _____

7. Set up a pair of simultaneous equations and use them to solve the following problem:

There are a total of 665 students attending a college. Of these the number of men M is 2½ as great as the number of women W. How many men students and how many women students are there?

_____ _____   _____

8. Solve the following set of equations for x, y, and z:

$$4x + 2y = 10$$
$$y - 6z = -15$$
$$x + 2z = 4$$

_____ _____ _____

ENDING TIME_____

ANSWERS:

1. 1120 mi

2. $\Delta L/L = k\,\Delta T$

3. Slope = 0.75; intercept = −1.

   $y = 0.75x - 1$

   _or_

   $4y = 3x - 4$

4. m = 0.38 kg/l (the density of the liquid)

   M_o = 0.75 kg (the tare weight of the beaker)

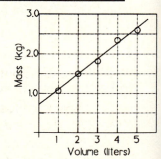

5. x = 2

6. 3 min 20 s

7. M= 475, W = 190

8. x = −8, y = 21, z = 6

ESSAY: Two Remarkable Proportions of Professor Newton

Space and time are the threads from which the fabric of all physical science is woven. Space is typically measured in units of distance strung out in three mutually perpendicular directions. Time units are correlated with the readings of a clock. But above all, physics is the study of change; and change has to do with motion, which requires combining the units of space and time. For instance, the measure of speed or velocity is always in terms of a ratio of distance to time units. (Velocity is nearly the same idea as speed, except that a statement of velocity also specifies the direction of the motion.)

Since so much of physics is clothed in ideas about motion and its causes, we trace the beginnings of modern science to the 17th-century scholars Galileo and Isaac Newton, who first clearly and correctly described the fundamental laws of motion. The first step was the recognition of the concept of inertia—the notion that it takes a push or a pull (a "force") to *change* the motion of a body. Otherwise, in the absence of a net force, a body in motion would continue to move in the same direction and with the same speed forever.

But what if the motion of some body is observed to change? How should the alteration in the motion be measured? And how is this to be related to forces which are responsible for the changes in the motion? The very simplest measure of modified motion is the rate of change of speed, i.e., the amount the speed is incremented divided by the time it takes for this to happen. This measure of change of motion is called *acceleration*. In symbols,

$$\text{acceleration, } a \equiv \frac{\text{speed increment}}{\text{time interval}} \equiv \frac{\Delta v}{\Delta t}$$

And the very simplest way the acceleration can be related to force is through a direct proportion; in symbols,

$$F \propto a .$$

This is Newton's famous 2nd law of motion; the constant of proportionality is called "mass" (symbolized by m), so that we usually see the 2nd law written

$$F = ma.$$

Mass is a measure of the amount of material in a body regarded in an "inertial" sense.

Newton was not merely interested in the motion of earth-bound objects. What of the motion of the moon about the earth? Or of the planets about the sun? To understand the motion of celestial bodies something must be known about the forces which act among them. In response to this need Newton asserted his law of universal gravitation which, among its other features, is also a direct proportion involving mass: the gravitational pull of any body of mass m on another body at a given distance is proportional to m. In symbols

$$F_{\text{gravity}} \propto m .$$

In terms of this law, mass is a measure of the amount of material in a body regarded in a "gravitational" sense.

For a given body how sure are we that mass regarded as a source of gravitational force is the same as the mass regarded as the source of inertia? In fact, excruciatingly careful measurements, beginning with those of the Hungarian physicist Eötvös in 1909, assure us that the two aspects of mass cannot differ by more than 1 part out of 10^{12}. Yet, to this day we cannot be sure if the quantities m used in each of Newton's two remarkable direct proportions are identical. It is unlikely that the apparent equality is just a great cosmic coincidence. More certainly it reflects a deeper truth about the meaning of matter which we have yet fully to understand.

Round III — Dealing with Space

Objects and events which are of concern to physics are described in terms of how and where they fit in space. So it is no wonder that the mathematics which deals with shape, size, and location is of such fundamental importance for successful problem solving. In this round, basic material from geometry and trigonometry with applications to physics-like problems is reviewed. You will also be getting more practice in visualizing problems and drawing diagrams. First try this pretest.

PRETEST—Optimum test time: 30 minutes or less.

Enter your answers and drawings in the right hand column, then check them against those given following the last question. A calculator may be used unless ruled out by the question. Keep track of your time.

STARTING TIME_____

ANSWERS

1. Make a sketch of the following: Equilateral triangle ABC has perpendiculars dropped from vertices A and C to the sides opposite. Indicate the values in degrees of all the angles formed by the lines in this figure.

2. Angles ABO and DCO are equal. How do you know triangles I and II are similar? How long is side OC?

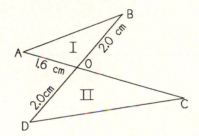

_____

_____

3. A square 1.0 cm on a side has a circle inscribed in it which is tangent to each of the sides. Draw a diagram. What is the area of each of the corner segments?

_____

4. What is the volume of the earth in cubic miles? Assume a spherical shape with a diameter of 7.9×10^3 miles.

_____

5. In the two right triangles shown below, some of the angles and side lengths (in arbitrary units) are given. What are the values of the remaining angles and side lengths?

(a)

(b)

(a) a = _____
b = _____
α = _____
(b) c = _____
α = _____
β = _____

6. You are standing on flat ground 150 feet from the base of a tree. A line drawn from your feet to the top of the tree subtends an angle of 35° with the horizontal. Make a sketch. How tall is the tree? _____

7. Complete each of the following statements by selecting from the given choices; do not use a calculator:

> CHOICES: sin 30°   cos 30°   tan 30°
>   −sin 30°   −cos 30°   −tan 30°

(a) sin(−30°) is equal to ...
(b) cos 330° is equal to ...
(c) tan 210° is equal to ...
(d) sin 150° is equal to ...

_____
_____
_____
_____

8. A parallelogram has sides of 15 cm and 24 cm which form an angle of 60°. Determine the length of the shorter diagonal and the angles it makes with each side. _____
_____

ENDING TIME_____

ANSWERS:

1.

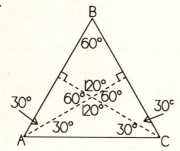

2. All corresponding angles are equal.
2.5 cm

3.

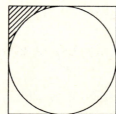

Area = 0.66 cm²

4. 2.6×10^{11} mi³

5. (a)   a = 1.7
          b = 1.0
          α = 60°
   (b)   c = 9.2
          α = 41°
          β = 49°

6.

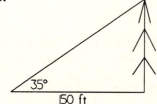

105 ft

7. (a)   −sin 30°
   (b)   cos 30°
   (c)   tan 30°
   (d)   sin 30°

8. 21 cm
   38°, 82°

Review 8 — Geometry I: Describing and Drawing

Geometry literally means "measuring the earth", but it has come to mean much more than ways to calculate the area of a farmer's field. This venerable subject is applicable even to the most modern problems. Setting up problems for solution often demands both a familiarity with the nomenclature of geometry and an ability to readily translate geometrical description into useful drawings. The basic ideas required are covered in this review.

PLANE FIGURES

Two dimensions. Although the objects of physics are obviously located in three-dimensional space of up-down, front-back, and side-to-side, much of what goes on is confined to two dimensions: the cables supporting an object from a derrick lie in a plane; planets move in flat orbits. Moreover, since the drawings we make are forced into the two dimensions of a sheet of paper or a chalkboard we need to think easily about the silhouettes and cross-sections which represent a problem. The common plane figures described in this section recur over and over in physics problems.

Triangles. The three straight lines forming this simplest of polygons must by necessity all lie in the same plane. Each corner of the triangle is called a *vertex*. It is commonplace to label the three angles by Greek letters and the sides opposite those angles by corresponding Roman letters.

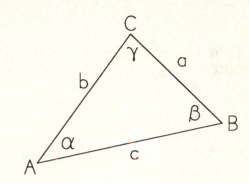

Special triangles. Of particular importance in physics problems are *right* triangles, *isosceles* triangles, and *equilateral* triangles. An example of each is shown here.

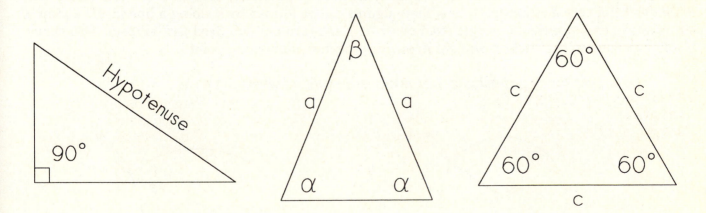

Basic properties of these triangles are as follows:

- *A right triangle contains a <u>right</u> angle (90°) opposite the side labeled the <u>hypotenuse</u>.*

• *An isosceles triangle has two equal angles (α). These are opposite two equal sides (a) which converge at the <u>apex</u> of the triangle. The side opposite the apex angle β is the <u>base</u> of the isosceles triangle.*

• *An equilateral triangle is completely symmetric. All sides are of equal length and all angles are equal to 60°.*

Four sided plane figures (quadrilaterals). Of these, we mostly need to know about regular figures with pairs of parallel sides. Shown here are a *parallelogram*, a *rectangle*, and a *square*:

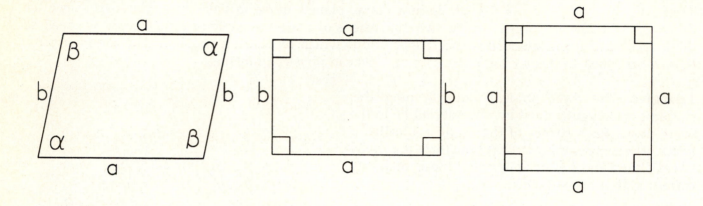

• *A parallelogram has pairs of equal angles (α and β) and opposite pairs of equal sides (a and b). Opposite sides are parallel.*

• *A rectangle is a parallelogram with all right angles.*

• *A square is a rectangle having four equal sides.*

Circles and ellipses. Circles crop up frequently in physics problems; ellipses occur less often. The drawings below are a reminder of how these figures can be formed by scribing a line inside a loop of string stretched out around a single fixed *center* (circle) or around two fixed *foci* (ellipse). Also shown is a line tangent to the circle, a straight line which touches at only one point.

• *A tangent line is perpendicular to a radius at the point of contact.*

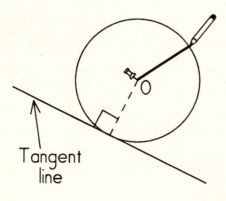

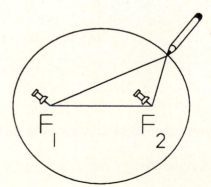

Making useful sketches. Some of the previous Reviews have stressed the importance of sketch-making in solving problems. The language and facts of geometry facilitate the art of making truly useful drawings. A ruler, compass, and protractor are usually not necessary; a reasonably careful freehand sketch is good enough most of the time.

  • *A "careful" drawing is not a masterpiece; but it should not mislead.*

For instance, when a circle is specified it should not look a lot like an ellipse; by the same token if an angle is *not* specified it should not appear to be a right angle. Simple aids to the memory are often used, such as an arrow to represent a direction of motion. Here are a couple of examples of word descriptions translated into serviceable drawings.

Make sketches representing the following situations:

  (a) A circular hoop rolling down a plane inclined at 30° to the horizontal; and

  (b) a square picture with sides of length a supported evenly from two corners by a string of total length 3a passing over a hook in the wall.

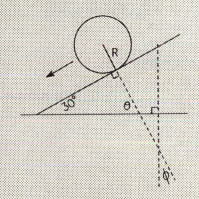
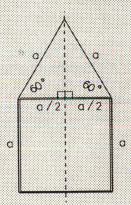

DISCUSSION: Some features which might be crucial to solving a problem have been included in the drawings for discussion.

  (a) The plane is tangent to the hoop, hence the plane is perpendicular to the radius R. The plane lies on the hypotenuse of a right triangle with vertical and horizontal sides. An extension of the radius meets the horizontal at angle θ and the vertical at angle φ.

  (b) The string and upper edge of the picture form an equilateral triangle with sides a. A vertical line through the hook divides this triangle into right triangles and divides the picture into rectangles with sides a and ½ a.

SOLID FIGURES

Important shapes. At the beginning level in physics there are a few 3-dimensional shapes which the student has to think about. Some common ones are the cube, rectangular parallelepiped, sphere, and right circular cylinder. To keep an illusion of three dimensions they can be sketched "in perspective" as shown here.

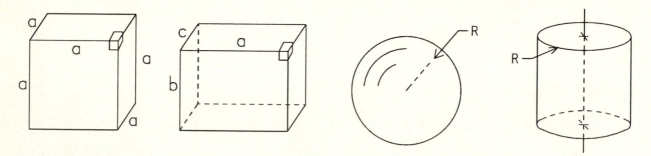

Here are some useful things to remember about these objects:

The rectangular parallelepiped is a figure formed from 6 rectangles joined by their edges; all the edges join at right angles. A cube is a rectangular parallelepiped formed from identical squares. The right circular cylinder can be thought of either as being generated by sliding a circle along its axis, or by wrapping an originally flat sheet around two identical parallel circular disks.

Drawing solid shapes. Notice some of the special touches which make the drawings above "believable." Right angle symbols emphasize the relationship of the lines at each corner of the cube and parallelepiped; dashed lines are used for hidden edges. Notice how a few contour lines can be used to distinguish a sphere from a circle. The "end caps" of the cylinder are drawn elliptical, which is how a circle appears when viewed off-axis.

With a little practice you should not find it difficult to make rough approximations of these drawings for yourself. Most problems in physics will probably not require such pictures, but occasionally making a "3-dimensional" perspective sketch can stimulate your thinking about a problem. Consider, for instance, the following example:

Identical atoms lie at each corner of a cube with sides of length a. Draw the line which joins two diagonally opposite atoms.

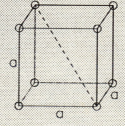

Skill Drill 8

In order to set up many physics problems for solution you should be able to readily translate word descriptions of a geometrical arrangement into a useable sketch. This entails knowing the nomenclature and making simple freehand drawings. These exercises provide practice in these basic skills.

1. Which of the following terms apply to each of the plane figures or combination of figures shown below? (More than one term may apply to a given figure.) Put the appropriate letter(s) in the answer spaces provided.

(a) right triangle	(e) square
(b) isosceles triangle	(f) circle
(c) equilateral triangle	(g) ellipse
(d) rectangle	(h) tangent line

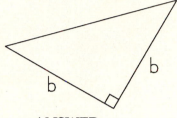

ANSWER_____

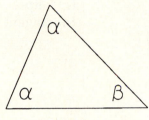

ANSWER_____

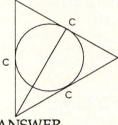

ANSWER_____

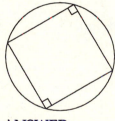

ANSWER_____

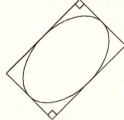

ANSWER_____

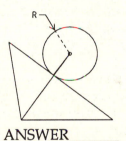

ANSWER_____

2. Beneath each of the following geometrical descriptions make a corresponding freehand sketch.

A right triangle with a vertically oriented hypotenuse

A circle inscribed in a square

Two isosceles triangles with a common base

A sphere tangent to a horizontal plane (seen edge on)

87

Skill Drill 8 — SOLUTIONS AND ANSWERS

1. Which of the following terms apply to each of the plane figures or combination of figures shown below? (More than one term may apply to a given figure.) Put the appropriate letter(s) in the answer spaces provided.

(a) right triangle
(b) isosceles triangle
(c) equilateral triangle
(d) rectangle

(e) square
(f) circle
(g) ellipse
(h) tangent line

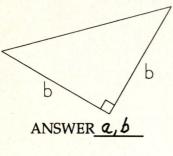

ANSWER _a, b_

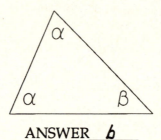

ANSWER _b_

ANSWER _a, c, f, h_

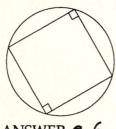

ANSWER _e, f_

ANSWER _d, g, h_

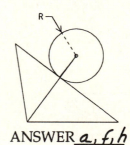

ANSWER _a, f, h_

2. Beneath each of the following geometrical descriptions make a corresponding freehand sketch.

A right triangle with a vertically oriented hypotenuse

A circle inscribed in a square

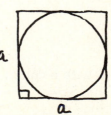

Two isosceles triangles with a common base

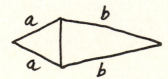

A sphere tangent to a horizontal plane (seen edge on)

A cube, showing three faces, each with a perpendicular line pointing out from the center of the face

An ellipse, with a tangent line at the end of the major axis

(3) In the space below each of the following descriptions make a freehand sketch which reasonably represents the situation. In most cases an edge-on view is sufficient.

A rectangular box sliding down an inclined plane surface

A horizontal sailboat boom being supported from the top of the mast by a rope

A slug of fluid moving through a cylindrical pipe

Two spheres colliding

Two blocks on either end of a seesaw

Rays of light directed radially outward from the sun

A rectangular chest being pulled up an incline by a rope

A descending yo-yo

A circular ring of wire surrounding a long straight wire located on its axis

A leaning slab of stone being supported by a spherical boulder

A rock falling from the leaning tower

A pendulum moving back and forth in a circular arc

A cube, showing three faces, each with a perpendicular line pointing out from the center of the face

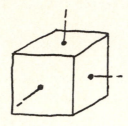

An ellipse, with a tangent line at the end of the major axis

(3) In the space below each of the following descriptions make a freehand sketch which reasonably represents the situation. In most cases an edge-on view is sufficient.

A rectangular box sliding down an inclined plane surface

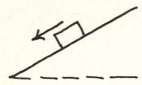

A horizontal sailboat boom being supported from the top of the mast by a rope

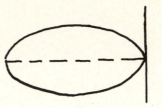

A slug of fluid moving through a cylindrical pipe

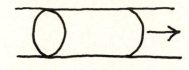

Two spheres colliding

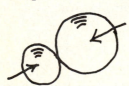

Two blocks on either end of a seesaw

Rays of light directed radially outward from the sun

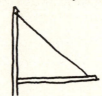

A rectangular chest being pulled up an incline by a rope

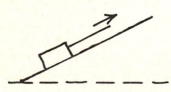

A descending yo-yo

A circular ring of wire surrounding a long straight wire located on its axis

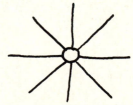

A leaning slab of stone being supported by a spherical boulder

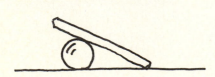

A rock falling from the leaning tower

A pendulum moving back and forth in a circular arc

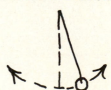

Review 9 — Geometry II: Angles, Shape, and Size

Making a sketch which shows the geometry of a problem is usually not sufficient for solving a problem. Quantitative information has to be used, but in many cases this is not given directly in the statement of the problem. Dimensions or angles which are needed to find an answer may have to be deduced from other lengths and angles shown in the diagram. The geometrical facts which can be used to find such relationships, as well as important formulas for areas and volumes, are reviewed in this section.

ANGLES AND POLYGONS

Angle measure. Angles are most commonly measured in degrees.   Radian measure, as an alternative for dealing with certain important types of problems, will be discussed in a later review.

Special names are given to angles which relate in a simple way to certain fractions of 360°, the angle swept out by a line making one complete rotation in a plane. Half a rotation results in a *straight angle* or 180°; a quarter rotation results in a *right angle* (90°). Any angle less than 90° is an *acute angle*. The "complement" of an acute angle is the difference between that angle and a straight angle. Any angle between 90° and 180° is called an *obtuse angle*.

Angles of a triangle. Angles in one part of a triangle can be related to other angles in that triangle using these facts:

> *(a) The sum of the <u>interior</u> angles of a triangle is a straight angle.*

In terms of the triangle pictured at the right $\alpha + \beta + \gamma = 180°$.

> *(b) An <u>exterior</u> angle equals the sum of the opposite interior angles.*

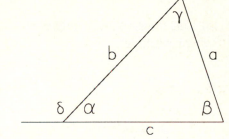

(An exterior angle is formed by extending one side of a triangle beyond the vertex.) In terms of the figure $\delta = \beta + \gamma$.

This problem applies these ideas to a drawing similar to one used in an example in the last Review:

A hoop is rolling down an inclined plane which makes an angle of 25° with respect to the horizontal.
>   (a) What angle does the incline make with respect to the vertical (α in this drawing)?
>   (b) When the hoop reaches the bottom of the incline, it runs out on a horizontal ramp. What angle does this ramp make with the incline (β in this drawing)?

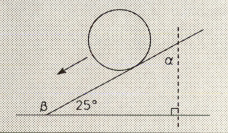

DISCUSSION: (a) In the right triangle shown, the three interior angles add up to a straight angle, i.e., $180° = \alpha + 25° + 90°$. Hence $\alpha = 65°$.

  (b) β is an exterior angle so that $\beta = \alpha + 90° = 155°$.

Angles in a quadrilateral. The following rule is easily understood when it is recognized that a diagonal line (dashed) drawn between opposite corners divides a four sided polygon into two triangles. Each of these triangles contributes 180° to the sum of the interior angles of the quadrilateral.

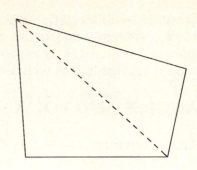

The interior angles of a plane quadrilateral add up to two straight angles.

SIMILAR TRIANGLES

Triangles which are of the same shape but differ only in size and/or orientation are called *similar triangles*. They often may arise in a drawing when triangles have parallel sides or common vertices. In the two examples shown here sides a and a' are parallel.

Three important properties of similar triangles are the following:

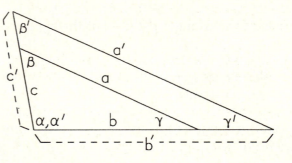

(a) *The three interior angles are the same in all similar triangles.*

In both pairs of triangles shown here $\alpha=\alpha'$, $\beta=\beta'$, and $\gamma=\gamma'$.

(b) *The three sides are in the same proportion in all similar triangles.*

In the examples of similar triangles shown here a:b:c = a':b':c'. A third property follows logically from this one:

(c) *All the sides of a triangle are in the same proportion to the corresponding sides of any triangle similar to it.*

In the triangles shown above a:a' = b:b'= c:c'.

Mirror image triangles. Some geometrical arrangements result in a pair of triangles whose shapes, while not identical, share all the characteristics of similar triangles just listed. This is the case when one of the triangles has the shape which the other would have were it reflected in a mirror. In the drawing the dashed lines show the mirror image of triangle ABC; triangle A'B'C', which has the same shape, can be considered to be similar to ABC. Be careful, however, in setting up proportions among the side lengths to recognize which correspond to one another.

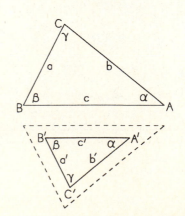

ANGLES FORMED BY INTERSECTING STRAIGHT LINES

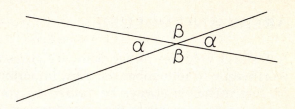

Opposing angles. Opposing angles are formed when two lines cross over each other. (In the drawing the two α's are opposing angles; so are the β's.)

> _Opposing_ angles formed by the intersection of two straight lines are equal.

This explains why $\alpha = \alpha'$ in the first example of similar triangles in the last section.

Angles formed by parallel lines. As shown in the drawing:

> _A straight line intersects parallel straight lines at equal angles._

This explains why sides a and a' are parallel in the first two cases of similar triangles shown on the previous page.

Angles formed by pairs of perpendiculars. In physics diagrams lines drawn perpendicular or "_normal_" to other lines frequently occur; for example, horizontal and vertical lines. An extremely useful fact is the following:

> _If a pair of lines intersect at some angle, two other lines which are perpendicular to them intersect at the same angle._

This example uses this idea to solve a problem involving similar triangles:

Referring to the last example of a hoop on an incline:

(a) What angle does the radius line which passes through the point of contact make with the vertical?

(b) If the hoop has rolled one-third of the way along the incline from the upper end, how much elevation above the lower horizontal ramp has it lost?

DISCUSSION: (a) Draw in a vertical (dashed line from point B) and extend the radius until it crosses it. (The acute angle of intersection is labelled θ.) Since the radius line is perpendicular to AB and the vertical line is perpendicular to AC, they intersect at the same angle which the incline makes with the horizontal. Thus θ = 25°.

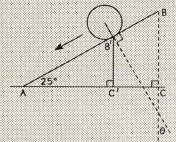

(b) Triangle AB'C' is similar to triangle ABC. Hence B'C' is in the same proportion to BC as AB' is to AB; one third of the original elevation has been lost.

AREAS AND VOLUMES

Areas and volumes of geometrical figures are needed often enough in physics problems so that it is worthwhile to remember the most important formulas. But, as elsewhere in physics, do not rely solely on rote memory; always be guided by your good sense. For instance all formulas for areas contain length measurements to the second power. Some formulas are simply extensions of other rules; several cases are noted below:

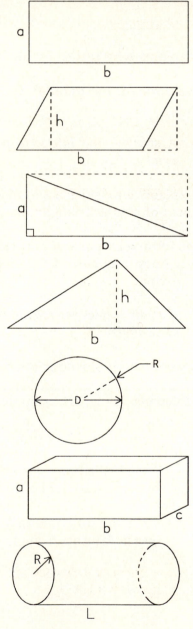

Rectangle.   *Area = ab* .

Parallelogram. A perpendicular of height h would cut off a triangle which would just fill in an area on the other side of the figure to form a rectangle; so

$$Area = hb.$$

Right triangle. This is half a rectangle; so

$$Area = ½ab .$$

Other triangles. A perpendicular of height h (the "altitude") divides the triangle into two right triangles whose areas combine to give simply

$$Area = ½hb.$$

Circles. Know the formulas both in terms of radius R and diameter D:

$$Area = \pi R^2 = \pi D^2/4$$
$$Circumference = 2\pi R = \pi D.$$

Rectangular parallelepipeds (and cubes). Surface area is the sum of the rectangular areas of all six faces. Also
$$Volume = abc .$$

Right circular cylinder. The surface area includes two "end caps" and the cylindrical "wrapper;" thus

$$Surface\ area = 2\pi R^2 + 2\pi RL.$$

The volume may be thought of as being generated by sweeping the cross-sectional area along the axis a distance L; thus

$$Volume = \pi R^2 L .$$

Sphere. A sphere has roughly half the volume and roughly half the surface area of a cube that would just hold it.
$$Volume = (4/3)\pi R^3 .$$
$$Surface\ area = 4\pi R^2 .$$

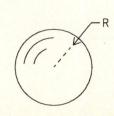

Skill Drill 9

The first problem is intended to remind you of major ideas in the preceding review. The next questions also involve abstract geometrical figures, but following that you are challenged to apply this understanding to some realistic situations.

1. Review of major points. (Fill in the blanks.)

(a) In triangle ABC $\alpha = 35°$ and $\beta = 70°$. Determine angle γ.

(b) Exterior angle θ is the complement of angle ___. What is θ in degrees?

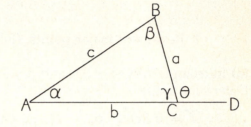

(c) Extend line BC to a point E so that DE is parallel to AB. Label the angles in triangle CDE which are equal to α, β, and γ.

(d) Fill in the blanks with the sides of triangle ABC (a,b,c) which are in the same proportion as the indicated sides of triangle CDE:

  CD:DE:EC = ___:___:___.

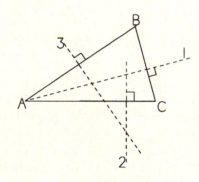

(e) Here triangle ABC has been redrawn with the addition of dashed lines 1, 2, and 3 perpendicular to each of the sides. At the points of intersection of the dashed lines label all angles which are equal to α, β, or γ.

(f) The distance along line 1 from vertex A to base BC is altitude h. Write down a formula for the area of triangle ABC in terms of h.

2. In the triangle shown here, find the remaining interior angles α and β:

3. A perpendicular to one of the sides has been added to the previous figure. Find θ and ϕ.

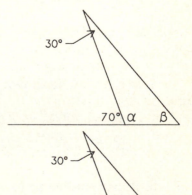

Skill Drill 9 — SOLUTIONS AND ANSWERS

1. Review of major points. (Fill in the blanks.)

(a) In triangle ABC α = 35° and β = 70°.
Determine angle γ.

$$\gamma = 180° - \alpha - \beta = 180° - 70 - 35° = 75°$$

(b) Exterior angle θ is the complement of angle $\underline{\gamma}$. What is
θ in degrees.

$$\theta = 180° - \gamma = \alpha + \beta = 105°$$

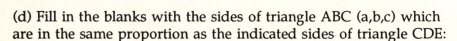

(c) Extend line BC to a point E so that DE is parallel to AB.
Label the angles in triangle CDE which are equal to α, β, and γ.

(d) Fill in the blanks with the sides of triangle ABC (a,b,c) which
are in the same proportion as the indicated sides of triangle CDE:

CD:DE:EC = $\underline{b:c:a}$. *The indicated sides as well as sides*
b, c, and a are opposite angles β, γ, and α, respectively.

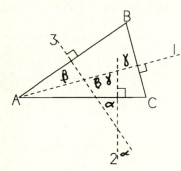

(e) Here triangle ABC has been redrawn with the addition of
dashed lines 1, 2, and 3 perpendicular to each of the sides. At the
points of intersection of the dashed lines label all angles which are
equal to α, β, or γ.

(f) The distance along line 1 from vertex A to base BC is altitude h.
Write down a formula for the area of triangle ABC in terms of h.

$$Area = \tfrac{1}{2} ha$$

2. In the triangle shown here, find the remaining interior angles
α and β:

$$\alpha = 180° - 70° = 110°$$
$$\beta = 180° - 110° - 30 = 40°$$

3. A perpendicular to one of the sides has been added to the
previous figure. Find θ and φ.

$$\theta = 180° - 90° - 40° = 50°$$
$$\phi = 180° - 30 = 60°$$

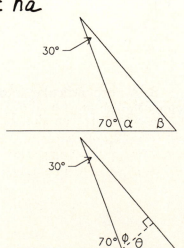

4. If $\phi = \phi'$, how do you know that triangles LNP and MNQ are similar? Given that LN = 0.7 m, NP = 0.3 m, and MN = 1.0 m, find NQ.

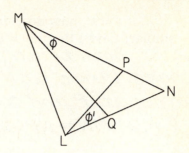

5. Side LM of triangle LMO, for which angles and side lengths are given in the figure, is parallel to side NP of triangle NOP. (a) Find values for the angles in triangle NOP. (b) If NP = 1.00 cm find the remaining sides of NOP.

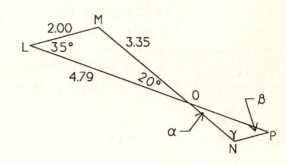

6. D is halfway between A and B, and E is halfway between C and B. (a) If AC = 1.0 m, how long is DE? (b) What are the angles α, β, δ, and γ shown in the figure?

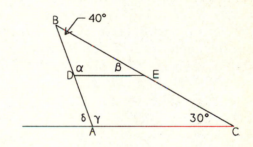

7. In the diagram at the right, find angles α, β, γ, and δ.

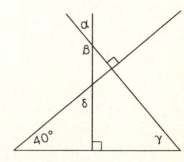

8. A rectangle 2 cm high and 5 cm wide has a triangular segment with sides 2 cm and 1 cm removed from one end. (a) What is the area of the remaining quadrilateral? (b) Remove another identical triangular segment from the other end. What is the area of the remaining parallelogram? (c) Check the last answer using a formula for the area of a parallelogram.

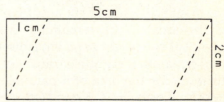

4. If $\phi = \phi'$, how do you know that triangles LNP and MNQ are similar? Given that LN = 0.7 m, NP = 0.3 m, and MN = 1.0 m, find NQ. *Both triangles have the same interior angles. (angle at N is common, thus labeled angles at P and Q are also equal.)*

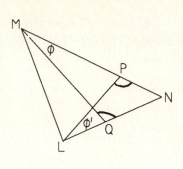

$$NQ : MN = NP : LN \quad \underline{or} \quad NQ/MN = NP/LN$$

$$Thus \ NQ = (1.0 \ cm) \left(\frac{0.3}{0.7}\right) = 0.4 \ cm$$

5. Side LM of triangle LMO, for which angles and side lengths are given in the figure, is parallel to side NP of triangle NOP. (a) Find values for the angles in triangle NOP. (b) If NP = 1.00 cm find the remaining sides of NOP.

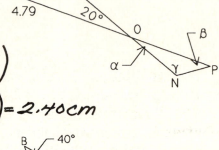

$$(a) \ \alpha = 20°, \ \beta = 35°, \ \gamma = 180° - \alpha - \beta$$
$$= 125°$$

$$(b) \ ON/NP = MO/LM \rightarrow ON = (1.00 \ cm) \left(\frac{3.35}{2.00}\right)$$
$$= 1.68 \ cm$$

$$OP/NP = OL/LM \rightarrow OP = (1.00 \ cm) \left(\frac{4.79}{2.00}\right) = 2.40 cm$$

6. D is halfway between A and B, and E is halfway between C and B. (a) If AC = 1.0 m, how long is DE? (b) What are the angles α, β, δ, and γ shown in the figure?

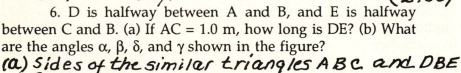

(a) Sides of the similar triangles ABC and DBE scale as 2:1. Thus
$$DE = 0.50 \ m$$

$$(b) \ \beta = 30°$$
$$\alpha = \gamma = 180° - 30° - 40° = 110°$$
$$\delta = 30° + 40° = 70°$$

7. In the diagram at the right, find angles α, β, γ, and δ.
Mutually perpendicular lines imply
$$\alpha = 40° \ and \ \delta = \gamma.$$
Then
$$\beta = 180° - \alpha = 140°$$
and
$$\gamma = \delta = 180° - 90° - 40° = 50°$$

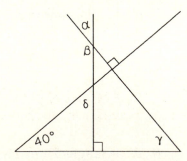

8. A rectangle 2 cm high and 5 cm wide has a triangular segment with sides 2 cm and 1 cm removed from one end. (a) What is the area of the remaining quadrilateral? (b) Remove another identical triangular segment from the other end. What is the area of the remaining parallelogram? (c) Check the last answer using a formula for the area of a parallelogram.

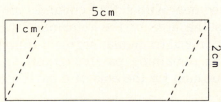

$$(a) \ Area = (5cm)(2cm) - \tfrac{1}{2}(1cm)(2 \ cm)$$
$$= 10 \ cm^2 - 1 \ cm^2 = 9 \ cm^2$$

$$(b) \ Area = 10 \ cm^2 - 2 \ cm^2 = 8 \ cm^2$$

$$(c) \ Area = hb = (2 \ cm)(4 \ cm) = 8 cm^2$$

9. Two lines are drawn from point P tangent
to the circle as shown. The line from the center O to
point P makes an angle of 30° with tangent line PT.

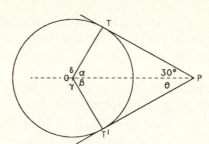

(a) What is the angle θ which the other tangent line
PT′ makes with the central axis?
(b) What are the angles (α, β, γ, δ) of the pie-shaped
segments formed by the radii to the tangent points?
(c) If the radius of the circle is 1.0 cm what are the areas of each pie shaped segment?

10. The drawing shows the parallel crests of a series of
waves approaching a beach in the direction shown by the arrow.
What is the angle ϕ which the crests make with respect to the
shoreline?

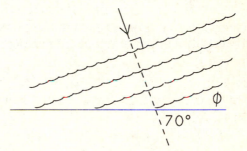

11. A straight-backed chair is tipped back so that the legs make
an angle of 65° with respect to the floor. What (acute) angles does the
seat make with respect to the floor? to the vertical?

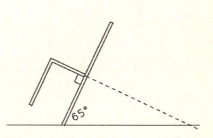

12. A bucket of paint is hanging by a piece of rope from the rung
of a ladder, which is leaning at an angle of 70° as shown.

(a) What angle α does the rope make with the ladder? (b) The arrow shows
the direction of the force with which the ladder presses on the wall. What
angle β does this force make with respect to the ladder?

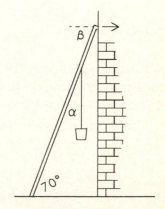

9. Two lines are drawn from point P tangent to the circle as shown. The line from the center O to point P makes an angle of 30° with tangent line PT.

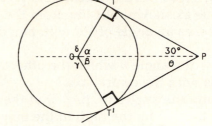

(a) What is the angle θ which the other tangent line PT′ makes with the central axis?
(b) What are the angles (α, β, γ, δ) of the pie-shaped segments formed by the radii to the tangent points?
(c) If the radius of the circle is 1.0 cm what are the areas of each pie shaped segment?

(a) Triangles OPT and OPT′ are identical; hence θ = 30°

(b) Tangent lines are perpendicular to radii; hence
α = β = 180° − 90° − 30° = 60°
Exterior angles γ = δ = 90° + 30° = 120° (Also note that δ and γ are complements of α and β.)

(c) Area of α and β segments = (60/360) πR² = 0.5 cm²
Area of γ and δ segments = (120/360) πR² = 1.0 cm²

10. The drawing shows the parallel crests of a series of waves approaching a beach in the direction shown by the arrow. What is the angle φ which the crests make with respect to the shoreline?

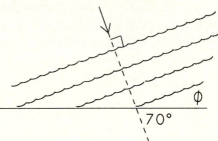

$$70° + φ = 90°$$
$$→ φ = 20°$$

11. A straight-backed chair is tipped back so that the legs make an angle of 65° with respect to the floor. What (acute) angles does the seat make with respect to the floor? to the vertical?

Label angles θ and φ, as shown.
θ = 180° − 90° − 65° = 25°.
The seat and vertical line are perpendicular to the lines forming θ. Hence φ = 25°

12. A bucket of paint is hanging by a piece of rope from the rung of a ladder, which is leaning at an angle of 70° as shown.

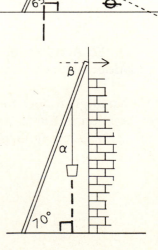

(a) What angle α does the rope make with the ladder? (b) The arrow shows the direction of the force with which the ladder presses on the wall. What angle β does this force make with respect to the ladder?

(a) α = 180° − 90° − 70° = 20°

(b) Ladder intersects parallel (horizontal) lines, hence
β = 70°

13. A box is being dragged up a 25° incline by a rope which makes an angle of 60° with respect to the horizontal. What is the angle θ which the rope makes with respect to the incline?

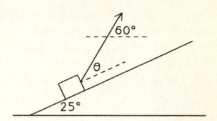

14. A long pole is held out at an angle from a building by a horizontal tie rope T attached to the top of the building. The point of attachment of the tie rope is 3/4 of the way from the bottom to the top of the pole. A weight W is suspended from the top of the pole by a rope so that W is in line with the top of the building.

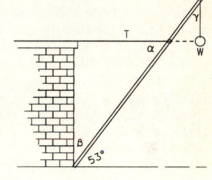

(a) If the pole is leaning at an angle of 53° find the angles α, β, and γ shown in the drawing. (b) If the building is 15 meters high, how long is the suspending rope? (c) If the tie rope is 12 meters long, how far out from the building is W?

15. What is the maximum volume of water which can be held in a cylindrical water tank 15.0 feet high and 10.0 feet in diameter?

16. Water stands in a cylindrical beaker 10 cm in diameter. If a marble 1.5 cm in diameter is dropped into it, how high does the water rise?

17. What is the surface area of the earth? Assume a spherical shape with a diameter of 7.9×10^3 miles.

18. How many miles does the earth travel in its orbit in a year, assuming a circular orbit of radius 93×10^6 miles? How many miles does it travel in a day?

13. A box is being dragged up a 25° incline by a rope which makes an angle of 60° with respect to the horizontal. What is the angle θ which the rope makes with respect to the incline?

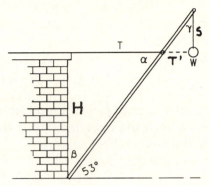

Draw in horizontal line, as shown.

$$\theta + 25° = 60°$$

Thus
$$\theta = 35°$$

14. A long pole is held out at an angle from a building by a horizontal tie rope T attached to the top of the building. The point of attachment of the tie rope is 3/4 of the way from the bottom to the top of the pole. A weight W is suspended from the top of the pole by a rope so that W is in line with the top of the building.

(a) If the pole is leaning at an angle of 53° find the angles α, β, and γ shown in the drawing. (b) If the building is 15 meters high, how long is the suspending rope? (c) If the tie rope is 12 meters long, how far out from the building is W?

(a) Tie rope is parallel to ground, thus
$$\alpha = 53°. \quad \beta = \gamma = 90° - 53° = 37°$$

(b) The two triangles shown are similar with ratio of sizes $\frac{3/4}{1/4} = 3/1$. Thus $S/H = 1/3 \rightarrow S = (15m)/3 = 5m$

(c) Likewise $T'/T = 1/3 \rightarrow T' = (12m)/3 = 4m$
Total distance from building $= 12m + 4m = 16m$

15. What is the maximum volume of water which can be held in a cylindrical water tank 15.0 feet high and 10.0 feet in diameter?

$$V = \pi R^2 L = \pi (5.0 ft)^2 (15.0 ft)$$
$$= 1180 ft^3$$

16. Water stands in a cylindrical beaker 5 cm in diameter. If a marble 1.5 cm in diameter is dropped into it, how high does the water rise?

Call rise in level Δh.

Volume of water displaced $(\pi D^2/4) \Delta h$
$$= \text{Volume of marble}, (4/3)\pi(d/2)^3$$
Solving $\Delta h = \frac{(4/3)\pi d^3/8}{\pi D^2/4} \quad \frac{2}{3} d^3/D^2 = 0.09\, cm$

17. What is the surface area of the earth? Assume a spherical shape with a diameter of 7.9×10^3 miles.

$$\text{Area} = 4\pi R^2 = 4\pi \left(\frac{7.9 \times 10^3\, mi}{2}\right)^2 = 2.0 \times 10^8\, mi^2$$

18. How many miles does the earth travel in its orbit in a year, assuming a circular orbit of radius 93×10^6 miles? How many miles does it travel in a day?

In a year, dist. $= 2\pi R = 2\pi (93 \times 10^6\, mi) = 5.8 \times 10^8\, mi$
In a day: $580 \times 10^6\, mi/365 = 1.6$ million miles

Review 10 — Trigonometry I: Mostly Right Triangles

Triangles occur not only in many problems of importance to physicists, but in those of the practical arts such as navigation and mechanical design. For example, a triangular structure formed by three rigid straight rods joined end-to-end has the special property that it cannot be squashed out of shape (as a rectangle can) without bending or breaking the rods. The vertex angles can only have certain specific values, depending on the lengths of the three sides of the triangle. Trigonometry deals with interrelationships among the sides and angles of triangles. In particular, right triangles merit special attention.

IMPORTANT SPECIAL TRIANGLES AND THEIR PROPERTIES

Since certain special shaped triangles occur repeatedly in physics problems their most important properties should be well understood.

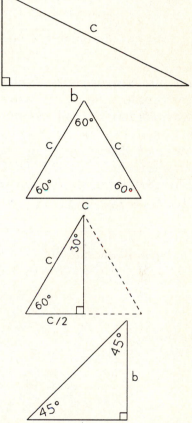

Right triangles—Pythagorean theorem. The lengths of the legs a and b are related to the length of the hypotenuse c according to

$$c^2 = a^2 + b^2 .$$

This gem-like rule was known by the Babylonians and proven to be exactly correct by the ancient Greeks. Yet it crops up even in the most advanced contexts in physics.

The equilateral triangle. This triangle is completely symmetric. All sides are equal in length. All angles are equal; each angle is 60°.

30–60–90 triangle. Split an equilateral triangle exactly in half. Each piece is a right triangle which is a favorite of writers of physics problems. The acute angles are 30° and 60°; the length of the side opposite the 30° angle is half that of the hypotenuse.

Right isosceles triangle. The acute interior angles are each 45°.

TRIGONOMETRIC FUNCTIONS

Apart from the useful facts about special triangles outlined above, trigonometry provides us with methods to relate sides and angles in triangles of any shape. However, instead of using values of angles themselves in its formulas, trigonometry relies on sets of numbers, called "trigonometric functions," which depend on values of angles.

Physics problems almost exclusively require only three trigonometric ("trig") functions in the solution of problems. These are the *sine*, the *cosine*, and the *tangent*. These functions can be defined in terms of the properties of a right triangle, as given in the next paragraph.

Sine, cosine, and tangent. Imagine acute angle θ at one vertex of a right triangle, as illustrated here. The hypotenuse and the legs of the triangle opposite and adjacent to θ have been labelled HYP, OPP, and ADJ respectively. The sine, cosine, and tangent of θ (abbreviated sin θ, cos θ, tan θ) are then defined by the following ratios:

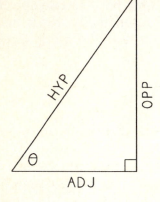

$$\sin \theta \equiv \frac{opposite\ side}{hypotenuse} \equiv \frac{OPP}{HYP}$$

$$\cos \theta \equiv \frac{adjacent\ side}{hypotenuse} \equiv \frac{ADJ}{HYP}$$

$$\tan \theta \equiv \frac{opposite\ side}{adjacent\ side} \equiv \frac{OPP}{ADJ}$$

Since the trigonometric functions can be defined as ratios of lengths, they have no units; they are dimensionless. Sine and cosine can never be greater in magnitude than 1, since the hypotenuse must always be longer than the other legs of a right triangle; however tangent can take on any value.

Values of sine, cosine, and tangent. Generally you can find values of trigonometric functions using your electronic calculator. (If you are using degrees set the DEG/RAD switch appropriately.) However, it is a worthwhile exercise to calculate (and ultimately, remember) the sine, cosine, and tangent of some of the more commonly occurring angles, as in this example:

Find the sine of 60°.

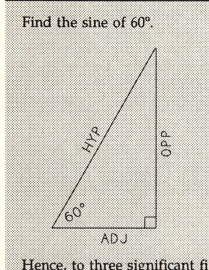

DISCUSSION: Recall that in the 30-60-90 triangle the shorter leg is half the hypotenuse. (This can be seen by realizing that this triangle is half an equilateral triangle.) Thus, replacing the labels shown in the drawing by
a = HYP and ½a = ADJ,
OPP can be obtained using the Pythagorean theorem as follows:

$$(OPP)^2 + (½a)^2 = a^2$$

$$OPP = (a^2 - a^2/4)^{½} = \sqrt{3}a/2$$

Hence, to three significant figures

$$\sin 60° = OPP/HYP = \sqrt{3}/2 = 0.867$$

Values of some other commonly occurring trigonometric functions, which can be calculated in a similar way (see Drill 10), are listed in the following table. You should realize, however, that although right triangles have been used to calculate the values shown in the table, and although the definitions of sine, cosine, and tangent have been stated in terms of right triangles, trigonometric functions can be used even in problems not involving right triangles.

TABLE — Trig functions of commonly used angles.

θ	sin θ	cos θ	tan θ
30°	1/2	√3/2	1/√3
60°	√3/2	1/2	√3
45°	1/√2	1/√2	1

USING TRIGONOMETRIC FUNCTIONS WITH RIGHT TRIANGLES

The most frequent way in which you will use trigonometry in physics problems is to find one side of a right triangular figure, given another side and an angle. For such problems it is useful to remember the meaning of sine and cosine in this form:

OPP = HYP sin θ   ADJ = HYP cos θ

This way of writing the definitions of sine and cosine is used in the first of these examples:

(a) A sailor steers his boat on a direct course 30° east of due north. After sailing 5.0 miles how far to the east and how far to the north has he travelled?

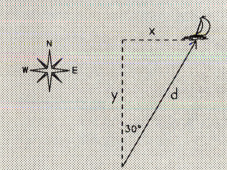

DISCUSSION: Make a diagram. Calling the distance along the direct course d, and the east and north distances x and y respectively, we have

x = d sin 30° = (5.0 mi)(½)
  = 2.5 miles.
y = d cos 30° = (5.0 mi)(√3/2)
  = 4.3 miles.

(b) On the edge of a river we see a tree directly across from us on the opposite side. Walking 50.0 yards along the edge of the river we again see the tree at an angle of 25° with respect to the river bank. How far away from our starting point is the tree?

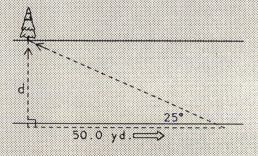

DISCUSSION: From this drawing we see that

tan 25° = d/(50.0 yd).

Solving for d and using a calculator to find tan 25° = 0.468, we have

d = (50.0 yd)(0.468) = 23.3 yd .

FINDING AN ANGLE GIVEN A TRIG FUNCTION

Inverse trig functions. In the examples given above an angle was the known quantity; the task in both cases was to calculate a side of a right triangle. The reverse problem is also important: given lengths of sides of a right triangle find one of the angles.

Suppose, for example, we know both the hypotenuse (HYP) and the side opposite the required angle θ (OPP), so that we can calculate the ratio $x = OPP/HYP \equiv \sin \theta$. The operation which takes us from x to its corresponding angle θ is called finding the "inverse sine" or the "arcsine" of x. This is written

$$\theta = \sin^{-1}x = \arcsin x.$$

This operation may be carried out using an electronic calculator by entering a value of x and pressing the appropriate function key. Likewise the calculator can be used to find the inverse cosine (arccosine)

$$\theta = \cos^{-1}y = \arccos y \ , \text{ where } y \text{ is the cosine of } \theta.$$

or the inverse tangent (arctangent)

$$\theta = \tan^{-1}z = \arctan z \ , \text{ where } z \text{ is the tangent of } \theta.$$

Application using a right triangle. Consider this example:

A car is driving along a straight inclined section of highway. When the mileage markers at the roadside indicate that it has travelled 1.00 mile the gain in elevation of the car is 500 ft. What is the angle of the incline with respect to the horizontal?

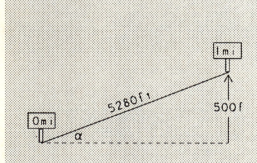

DISCUSSION: Referring to the drawing (not to scale), the angle we are asked to find is α. Thus, in terms of the usual notation for labelling a right triangle OPP = 500 ft and HYP = 1.00 mi = 5280 ft . Thus

$$\alpha = \arcsin(500/5280)$$
$$= \arcsin(0.0947)$$
$$= 5.43° .$$

Significant figures. In the example above the answer is rounded off to three significant figures. This is done because (as you can verify with a calculator) a change in the value of the sine (0.0947) by one unit in the last place affects the value of the arcsine by approximately 0.01°. In general, there is no simple way to determine the correct number of significant figures in an angle. But in most cases the context of the problem makes it clear how many places are reasonable.

Skill Drill 10

Following the questions which deal with abstract triangles, there are problems which apply trigonometry to concrete situations. A few problems resemble those given in other drills, but different questions are asked. In this drill, feel free to use a calculator whenever it is needed.

1. Review of major ideas—refer to the right triangle ABC shown here.

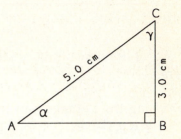

(a) Using the Pythagorean theorem find the length of side AB.

(b) Determine sin α, cos α, and tan α from ratios of side lengths.

(c) Find α and γ using inverse trig functions.

(d) Drop a perpendicular from vertex B to the hypotenuse AC. Find the length h of this line (an "altitude" of ABC) using a trig function of α.

NOTE: Triangle ABC in the problem above and all other triangles which are similar to it are the only possible right triangles whose side lengths form rational fractions (3/5, 3/4, 4/5).

2. Use the Pythagorean theorem to find cos 60° and tan 60°. Take the same approach used to find sin 60° in the example problem of Review 10.

3. Find the altitude h of an equilateral triangle whose sides a are each 1.0 cm in length. Do this using (a) the Pythagorean theorem, and (b) a trig function of a vertex angle.

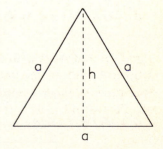

Skill Drill 10 — SOLUTIONS AND ANSWERS

1. Review of major ideas—refer to the right triangle ABC shown here.

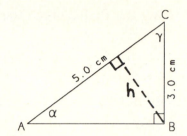

(a) Using the Pythagorean theorem find the length of side AB.

$$(AB)^2 = (5.0\,cm)^2 - (3.0\,cm)^2 = 16.0\,cm^2$$
$$\rightarrow AB = 4.0\ cm$$

(b) Determine sin α, cos α, and tan α from ratios of side lengths.

$$\sin α = 3.0/5.0 = 0.60$$
$$\cos α = 4.0/5.0 = 0.80$$
$$\tan α = 3.0/4.0 = 0.75$$

(c) Find α and γ using inverse trig functions.

$$α = \arcsin 0.60 = 36.9° \quad (\text{or } α = \arccos 0.80)$$
$$γ = \arccos 0.60 = 53.1°$$

(d) Drop a perpendicular from vertex B to the hypotenuse AC. Find the length h of this line (an "altitude" of ABC) using a trig function of α.

$$h = (AB) \sin α = (4.0\,cm)(0.60) = 2.4\,cm$$

NOTE: Triangle ABC in the problem above and all other triangles which are similar to it are the only possible right triangles whose side lengths form rational fractions (3/5, 3/4, 4/5).

2. Use the Pythagorean theorem to find cos 60° and tan 60°. Take the same approach used to find sin 60° in the example problem of Review 10.

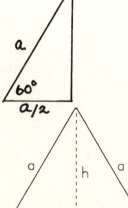

$$HYP = a, \quad ADJ = a/2 \rightarrow OPP = \sqrt{a^2 - (a/2)^2} = \sqrt{3}\,a/2$$
$$\text{Thus: } \cos 60° = ADJ/HYP = 1/2$$
$$\tan 60° = OPP/ADJ = \sqrt{3}$$

3. Find the altitude h of an equilateral triangle whose sides a are each 1.0 cm in length. Do this using (a) the Pythagorean theorem, and (b) a trig function of a vertex angle.

(a) Altitude line bisects base. Hence

$$h = \sqrt{(1.0\,cm)^2 - (0.5\,cm)^2} = 0.87\,cm$$

(b) All angles are 60°. Thus

$$h = (1.0\,cm) \sin 60° = 0.87\,cm$$

4. Apply the Pythagorean theorem to the isosceles right triangle shown here to find sin 45°, cos 45° and tan 45°.

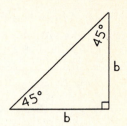

5. Use trig functions to determine the lengths of sides a and b (in arbitrary units) of the right triangles shown below.

(a)

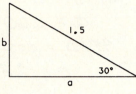

(b)

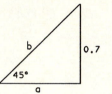

(c)

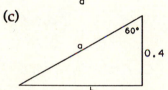

(d)

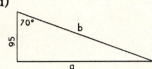

(e)

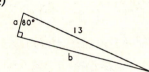

(f)

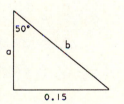

6. Use inverse trig functions to determine, to an accuracy of 0.1°, angles α and β in the right triangles whose side lengths (in arbitrary units) are shown in the figures below.

(a)

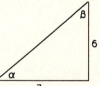

(b)

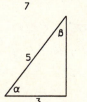

(c)

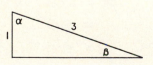

(d)

(e)

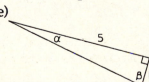

(f)

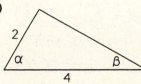

4. Apply the Pythagorean theorem to the isosceles right triangle shown here to find sin 45°, cos 45° and tan 45°.

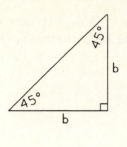

$$OPP = ADJ = b$$

$$HYP = \sqrt{b^2 + b^2} = \sqrt{2}\, b$$

$$Thus:\ \sin 45° = \cos 45° = \frac{b}{\sqrt{2}\,b} = \frac{1}{\sqrt{2}}$$

$$\tan 45° = \frac{b}{b} = 1$$

5. Use trig functions to determine the lengths of sides a and b (in arbitrary units) of the right triangles shown below.

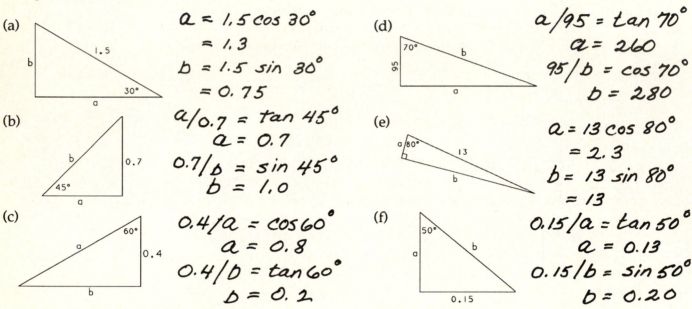

(a)

$$a = 1.5 \cos 30°$$
$$= 1.3$$
$$b = 1.5 \sin 30°$$
$$= 0.75$$

(b)

$$a/0.7 = \tan 45°$$
$$a = 0.7$$
$$0.7/b = \sin 45°$$
$$b = 1.0$$

(c)

$$0.4/a = \cos 60°$$
$$a = 0.8$$
$$0.4/b = \tan 60°$$
$$b = 0.2$$

(d)

$$a/95 = \tan 70°$$
$$a = 260$$
$$95/b = \cos 70°$$
$$b = 280$$

(e)

$$a = 13 \cos 80°$$
$$= 2.3$$
$$b = 13 \sin 80°$$
$$= 13$$

(f)

$$0.15/a = \tan 50°$$
$$a = 0.13$$
$$0.15/b = \sin 50°$$
$$b = 0.20$$

6. Use inverse trig functions to determine, to an accuracy of 0.1°, angles α and β in the right triangles whose side lengths (in arbitrary units) are shown in the figures below.

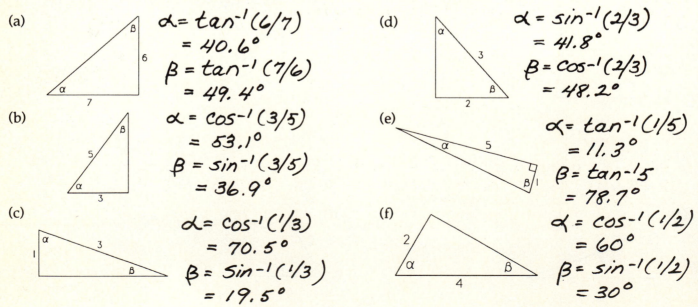

(a)

$$\alpha = \tan^{-1}(6/7)$$
$$= 40.6°$$
$$\beta = \tan^{-1}(7/6)$$
$$= 49.4°$$

(b)

$$\alpha = \cos^{-1}(3/5)$$
$$= 53.1°$$
$$\beta = \sin^{-1}(3/5)$$
$$= 36.9°$$

(c)

$$\alpha = \cos^{-1}(1/3)$$
$$= 70.5°$$
$$\beta = \sin^{-1}(1/3)$$
$$= 19.5°$$

(d)

$$\alpha = \sin^{-1}(2/3)$$
$$= 41.8°$$
$$\beta = \cos^{-1}(2/3)$$
$$= 48.2°$$

(e)

$$\alpha = \tan^{-1}(1/5)$$
$$= 11.3°$$
$$\beta = \tan^{-1}5$$
$$= 78.7°$$

(f)

$$\alpha = \cos^{-1}(1/2)$$
$$= 60°$$
$$\beta = \sin^{-1}(1/2)$$
$$= 30°$$

7. A heavy load is supported from the end of a boom, 6.0 m in length, held out from the mast of a derrick by a guy wire, as shown. If the boom makes a 45° angle with respect to the mast, how far from the mast is the end of the boom?

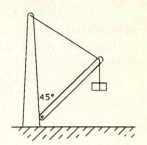

8. A flag pole is held out from the side of a building by a horizontal tie rope attached 16.0 feet above the bottom of the pole, as shown. (a) If the tie rope is 8.0 feet in length, what is angle α? (b) Use a trig function of α to find how far from the building the top of the pole is, assuming it is 22.0 feet long.

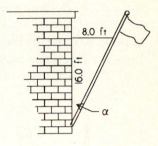

9. A box is being pulled up a 25° incline. After it has moved 3.0 meters along the incline, what is the distance x it has moved in the horizontal direction and what elevation y has it achieved?

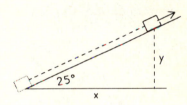

10. A child is swinging on a garden swing with supporting rope lengths of 10.0 feet. When the swing angle (with respect to the vertical) is 30°, how high is the child compared with her lowest position?

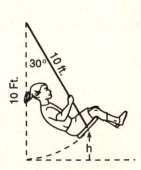

11. A stairway is built so that each step is 8 inches higher and is set back 10 inches from the next lower step. At what angle θ with respect to the horizontal does the stairway rise?

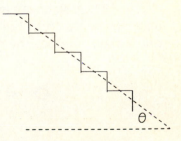

7. A heavy load is supported from the end of a boom, 6.0 m in length, held out from the mast of a derrick by a guy wire, as shown. If the boom makes a 45° angle with respect to the mast, how far from the mast is the end of the boom?

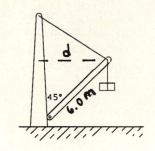

$$\text{Distance } d = (6.0m) \sin 45°$$
$$= 6.0m / \sqrt{2}$$
$$= 4.2\, m$$

8. A flag pole is held out from the side of a building by a horizontal tie rope attached 16.0 feet above the bottom of the pole, as shown. (a) If the tie rope is 8.0 feet in length, what is angle α? (b) Use a trig function of α to find how far from the building the top of the pole is, assuming it is 22.0 feet long.

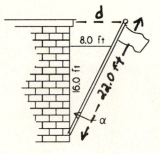

$$(a)\ \alpha = \tan^{-1}(8.0/16.0)$$
$$= 26.6°$$
$$(b)\ \text{Distance } d = (22.0ft) \sin \alpha$$
$$= 9.8\, ft$$

9. A box is being pulled up a 25° incline. After it has moved 3.0 meters along the incline, what is the distance x it has moved in the horizontal direction and what elevation y has it achieved?

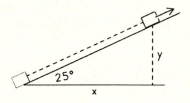

$$X = (3.0\, m) \cos 25° = 2.7\, m$$
$$y = (3.0\, m) \sin 25° = 1.3\, m$$

10. A child is swinging on a garden swing with supporting rope lengths of 10.0 feet. When the swing angle (with respect to the vertical) is 30°, how high is the child compared with her lowest position?

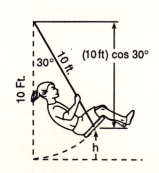

$$h = 10.0\, ft - 10.0\, ft \cos 30°$$
$$= 1.3\, ft$$

11. A stairway is built so that each step is 8 inches higher and is set back 10 inches from the next lower step. At what angle θ with respect to the horizontal does the stairway rise?

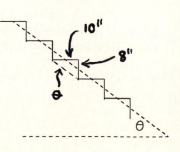

$$\theta = \arctan(8/10) = 39°$$

Review 11 — Trigonometry II: Other Triangles and Applications

The examples given in the previous review used right triangles only. In this section, triangles of a general shape are considered, as well as the idea of trigonometric functions of angles which are not positive or acute. Force, as a concept for which trigonometry is useful, is introduced.

TRIG FUNCTIONS OF NON-ACUTE ANGLES

Quadrants and angle convention. It is convenient to picture the entire range of angles as being generated by a straight line sweeping around through the four quadrants of a plane pictured at the right. By convention

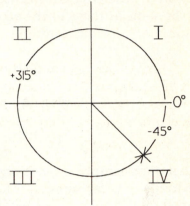

* *angles from 0° to 360° are measured <u>counterclockwise</u> starting from the positive horizontal axis, and*

* *angles from 0° to -360° are measured <u>clockwise</u> starting from the same axis.*

Thus angles between 0° and 90° (acute angles) are said to be in the "first quadrant" or Quadrant I, angles between 90° and 180° are in Quadrant II, etc.

Any given direction can be specified by either a negative or a positive angle. Negative angles may sometimes be a simpler choice, especially in Quadrant IV. For example, the orientation of a line which is 45° *below* the horizontal positive axis may be described either by θ = -45° or by θ = +315°.

Sine, cosine, tangent. The sine, cosine, and tangent of angles in Quadrants II through IV are defined much like they are defined for angles in the first quadrant. However various trig functions may, in certain quadrants, take on negative values. Here is how the trig functions can be defined in general:

> *1) Draw a line radially out from the origin pointing in the required direction, then form a right triangle by dropping a perpendicular from the end of the line to the horizontal axis.*

As an example, the drawing shows a right triangle formed from a line drawn at an angle of 120°.

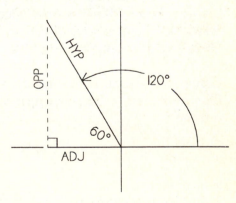

> *2) Except for sign each trig function is defined exactly as before in terms of this triangle, with the vertical side playing the role of OPP, the horizontal side playing the role of ADJ, and the radial line playing the role of HYP.*

The labels OPP and ADJ shown in the drawing are appropriate since these sides are opposite and adjacent, respectively, to the vertex at which the 120° angle is located. However, these quantities can take on negative as well as positive values, as follows:

3) *The sign of the trig function is determined by assigning a positive value to the vertical side (OPP) if it is in the upper half of the plane, and a negative value if it is in the lower half; likewise ADJ is assigned a positive or negative value depending on whether it is in the right half of the plane (+) or in the left half (-). HYP is always regarded as positive.*

Thus, in the example of the trig functions of 120° shown in the drawing, the sine is positive since OPP is positive, whereas cosine is negative since ADJ is negative. The size ("magnitude") of the trig functions are the same as those of the acute angle (60°) at the vertex of the triangle in the center of the drawing. In summary, for this example:

$$\sin 120° = \text{OPP}/\text{HYP} = \sin 60° = \sqrt{3}/2$$

$$\cos 120° = \text{ADJ}/\text{HYP} = -\cos 60° = -1/2$$

$$\tan 120° = \text{OPP}/\text{ADJ} = -\tan 60° = -\sqrt{3}$$

Rather than simply *memorizing* rules about the signs and magnitudes of trig functions in various quadrants it is far more satisfactory to quickly sketch the angle and examine the triangle which is formed by dropping a perpendicular to the axis.

Ambiguity in finding inverse trig functions. The example discussed above illustrates the fact that there is more than one angle corresponding to a given value of a trig function. An electronic calculator returns angle values only between −90° and +90° (Quadrants I and IV) when the $\sin^{-1}$, $\cos^{-1}$, and $\tan^{-1}$ keys are used, even though in an occasional problem the desired angle may not lie in the indicated quadrant. Invariably, however, a sketch accompanying the problem will clarify what the correct answer should be—just another reason why a sketch is important in problem solving!

"NON-RIGHT" TRIANGLES

The law of cosines. In an introductory physics course most of the trigonometry which is needed involves right triangles only. Occasionally, however, it may be easier to use one of two rules which relate side lengths and angles in triangles of a general type. The first of these rules, the *law of cosines*, gives a relationship among three sides of a triangle and one of the angles, as follows:

Using the usual notation for sides and corresponding angles, as illustrated here, when the angle involved is α, the law of cosines can be written

$$a^2 = b^2 + c^2 - 2bc \cos \alpha .$$

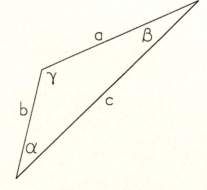

When either of the other angles is involved the corresponding statements of the law are

$$b^2 = c^2 + a^2 - 2ca \cos \beta$$

or

$$c^2 = a^2 + b^2 - 2ab \cos \gamma .$$

The law of cosines is not difficult to remember; it is like the Pythagorean theorem, except for the cross term involving two side lengths and the cosine of the included angle.

This example problem makes use of the law of cosines.

Mr. Smith strolls out into the woods from his fishing camp. He walks a half mile due north, then makes a 150° right turn and walks another 3/4 mile in a straight line. How far is he from his camp?

DISCUSSION: First draw a diagram. The two legs of his route and the line joining his final position and the camp form a general triangle with angle $\alpha = 30°$ at the corner where he changed directions. The distance to the fishing camp is given by

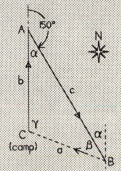

$$a^2 = b^2 + c^2 - 2bc \cos 30°$$

$$= (0.50 \text{ mi})^2 + (0.75 \text{ mi})^2$$
$$- 2(0.50 \text{ mi})(0.75 \text{ mi})(0.866).$$

Distance to camp, $a = 0.40$ miles.

In the above example, since a, b, and c are all finally determined, the law of cosines can again be used (in the appropriate form) to solve for each of the unknown angles β and γ. Instead, however, we will find these angles in a follow-up example using another useful rule: the "law of sines."

The law of sines. This rule expresses the remarkable fact that there is a simple direct proportionality between any side of a triangle and the sine of its opposing angle. Using our usual notation for angles and sides of triangles, the law of sines can be stated:

$$\frac{a}{\sin \alpha} = \frac{b}{\sin \beta} = \frac{c}{\sin \gamma}$$

The following extension of the previous example uses this rule.

In what direction from his present position (point B in the previous diagram) does Mr. Smith's camp lie?

DISCUSSION: Compass directions are usually stated with respect to the north-south line (dashed line in the drawing). The direction to the camp is thus $\alpha + \beta$. (AC is parallel to the N-S line; hence AB forms an angle α with both these lines.) To find β use the law of sines, as follows:

$$\frac{\sin \beta}{0.50 \text{ mi}} = \frac{\sin \alpha}{0.40 \text{ mi}}, \quad \text{where } \sin \alpha = \sin 30° = 0.500 .$$

$$\sin \beta = 0.625, \quad \text{or } \beta = \sin^{-1}(0.625) = 39°.$$

Hence the direction to the camp is $\alpha + \beta = 30° + 39° = 69°$ west of north.

FORCES — ANOTHER APPLICATION FOR TRIGONOMETRY

Forces quantitatively represented by arrows. The direction of a force (a push or a pull) can be depicted in the diagram of a problem by an arrow pointing in the appropriate direction. In addition, in quantitative problems it is useful to draw the length of force arrows in proportion to the strengths of the forces they represent, just as lengths of arrows on a map represent distances travelled. In fact, we can use geometry and trigonometry to analyze the effects of forces in the same way in which certain aspects of motion are analyzed.

Component forces. In some of the examples given in this review and the last one, motion of an object was regarded as a combination of separate motions along different directions. For instance, the track of a ship can be described in terms of the distances the ship moves towards the east and towards the north. In other words, the motion is a combination of an eastward "component" and a northward "component." On a map, these distances are the sides of a right triangle lined up along the east and north directions (d_E and d_N, respectively, in this drawing).

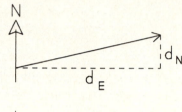

Likewise, it is often fruitful to regard a given force as a combination of component forces—for example, an "x-component" and a "y-component." To find these, the force arrow is drawn as the hypotenuse of a right triangle; the lengths of the sides parallel to the x- and y-directions are proportional to the x- and y-components, respectively. This is illustrated at the right by a force arrow representing a 10 lb force pointing in the designated direction. The components are

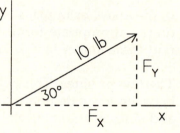

$$F_x = (10 \text{ lb})\cos 30° = 8.7 \text{ lb}$$
$$F_y = (10 \text{ lb})\sin 30° = 5.0 \text{ lb} .$$

Skill Drill 11

This drill provides practice using triangles of a general shape and angles in various quadrants. As in the previous drill, a calculator may be used when appropriate.

    1. Review of major points:

(a) What is the quadrant and equivalent negative angle of 255°? Label the quadrant and the angles in a diagram showing a line pointing in the appropriate direction.

(b) What acute angle has the same magnitude (absolute size) trig functions as 255°? Show this angle in your drawing.

(c) What are the *signs* of the sine, cosine, and tangent of 255°? (Do not use a calculator.)

(d) A triangle is formed using side lengths (in arbitrary units) a = 2, b = 3, and c = 4. Use the law of cosines to determine the angle γ formed by the two shorter sides.

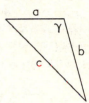

(e) Use the law of sines to find the remaining two angles of this triangle.

    2. In the space beneath each of the following angles, sketch a pair of x-y axes and a line pointing approximately in the indicated direction. Also label and give the value of an equivalent angle having an opposite sign. (Use the usual convention for angle measurement, i.e., positive angles are measured counterclockwise from the +x axis.)

(a) −67°   (b) −170°   (c) 340°   (d) −200°

    3. Determine the sine, cosine, and tangent of three of the angles listed in the previous question by finding the trig functions of an appropriate positive acute angle and assigning the proper sign.

(a) −67°   (b) −170°   (c) 340°

Skill Drill 11 — SOLUTIONS AND ANSWERS

1. Review of major points:

(a) What is the quadrant and equivalent negative angle of 255°? Label the quadrant and the angles in a diagram showing a line pointing in the appropriate direction.   *Quadrant III and (−105°)*

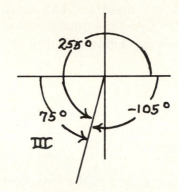

(b) What acute angle has the same magnitude (absolute size) trig functions as 255°? Show this angle in your drawing.   *75°.*

(c) What are the *signs* of the sine, cosine, and tangent of 255°? (Do not use a calculator.)   *Both OPP and ADJ are negative, hence: sine is neg., cosine is neg., tangent is positive.*

(d) A triangle is formed using side lengths (in arbitrary units) a = 2, b = 3, and c = 4. Use the law of cosines to determine the angle γ formed by the two shorter sides.

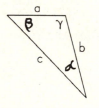

$$c^2 = a^2 + b^2 - 2ab\cos\gamma$$

Thus: $\cos\gamma = -(16-4-9)/12 = -0.25$ (Quadrant II)

and $\gamma = \cos^{-1}(-0.25) = 104°$

(e) Use the law of sines to find the remaining two angles of this triangle.

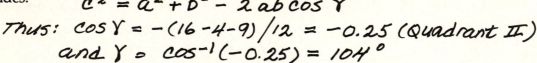

$$\sin\alpha = a\,\frac{\sin\gamma}{c} = 2\left(\frac{0.970}{4}\right) = 0.485$$

$$\alpha = 29°$$

$$\sin\beta = b\,\frac{\sin\alpha}{a} = 3\left(\frac{0.485}{2}\right) = 0.727$$

$$\beta = 47°$$

2. In the space beneath each of the following angles, sketch a pair of x-y axes and a line pointing approximately in the indicated direction. Also label and give the value of an equivalent angle having an opposite sign. (Use the usual convention for angle measurement, i.e., positive angles are measured counterclockwise from the +x axis.)

(a) −67°   (b) −170°   (c) 340°   (d) −200°

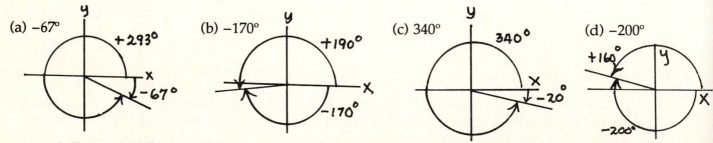

3. Determine the sine, cosine, and tangent of three of the angles listed in the previous question by finding the trig functions of an appropriate positive acute angle and assigning the proper sign.

(a) $-67° \equiv -\theta$

$$\sin(-\theta) = -\sin 67° = -0.92$$
$$\cos(-\theta) = \cos 67° = 0.39$$
$$\tan(-\theta) = -\tan 67° = 2.36$$

(b) $-170° \equiv -\theta$

$$\sin(-\theta) = \sin 10° = -0.17$$
$$\cos(-\theta) = -\cos 10° = -0.98$$
$$\tan(-\theta) = \tan 10° = 0.18$$

(c) $340° \equiv \theta$

$$\sin\theta = -\sin 20° = 0.34$$
$$\cos\theta = \cos 20° = 0.94$$
$$\tan\theta = -\tan 20° = -0.36$$

4. Use the law of cosines, the law of sines, or a combination of these laws to find the remaining angles and sides (in arbitrary units) of the triangles with the following measures. You may also use the geometrical fact $\alpha + \beta + \gamma = 180°$. Follow the usual convention for naming sides and angles (α opposite a, etc.)

(a) a = 5.0, c = 4.0,
$\beta = 50°$

(b) a = 7.0, b = 15.0,
c = 18.0

(c) c = 600, $\alpha = 65°$,
$\beta = 70°$

5. An airplane flies due north for 100 miles. It then makes a left turn directly to the northwest and flies another 100 miles. (a) What is the direct line distance to the starting point? (b) At what angle with respect to south does the starting point lie?

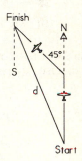

6. A tightrope walker is balanced on a wire strung between two buildings 100 feet apart. The wire dips at angles of 5° and 10° at the support points, as shown. What are the lengths of the segments of wire between where he is standing and each building? (Hint: Use law of sines.)

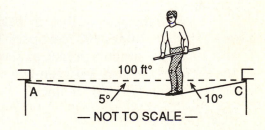

7. A tugboat is moving a raft of logs by pulling on a tow rope with a force of 1600 lb at an angle of θ with respect to east. This force has a component along the N-S direction and a component along the E-W direction. Determine the size (in pounds) of each component and state whether its direction is N, S, E, or W, for the following values of θ.

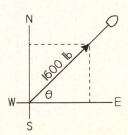

(a) $\theta = 60°$

(b) $\theta = 120°$.

4. Use the law of cosines, the law of sines, or a combination of these laws to find the remaining angles and sides (in arbitrary units) of the triangles with the following measures. You may also use the geometrical fact $\alpha + \beta + \gamma = 180°$. Follow the usual convention for naming sides and angles (α opposite a, etc.)

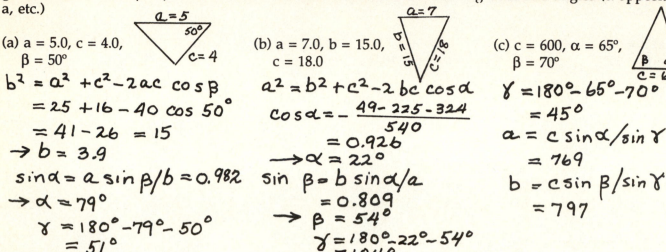

(a) a = 5.0, c = 4.0, $\beta = 50°$

$b^2 = a^2 + c^2 - 2ac \cos\beta$
$= 25 + 16 - 40 \cos 50°$
$= 41 - 26 = 15$
$\rightarrow b = 3.9$
$\sin\alpha = a \sin\beta/b \approx 0.982$
$\rightarrow \alpha = 79°$
$\gamma = 180° - 79° - 50°$
$= 51°$

(b) a = 7.0, b = 15.0, c = 18.0

$a^2 = b^2 + c^2 - 2bc \cos\alpha$
$\cos\alpha = -\dfrac{49 - 225 - 324}{540}$
$= 0.926$
$\rightarrow \alpha = 22°$
$\sin\beta = b \sin\alpha/a$
$= 0.809$
$\rightarrow \beta = 54°$
$\gamma = 180° - 22° - 54°$
$= 104°$

(c) c = 600, $\alpha = 65°$, $\beta = 70°$

$\gamma = 180° - 65° - 70°$
$= 45°$
$a = c \sin\alpha/\sin\gamma$
$= 769$
$b = c \sin\beta/\sin\gamma$
$= 797$

5. An airplane flies due north for 100 miles. It then makes a left turn directly to the northwest and flies another 100 miles. (a) What is the direct line distance to the starting point? (b) At what angle with respect to south does the starting point lie?

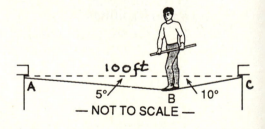

(a) $d^2 = (100\ mi)^2 + (100\ mi)^2 - 2(100\ mi)^2 \cos 135°$
$= (100\ mi)^2 [1 + 1 + 2(-0.707)]$
$= (100\ mi)^2 (3.41) \longrightarrow d = 185\ mi$

(b) $\sin\theta = (100\ mi) \sin 135°/185\ mi = 0.382$
$\longrightarrow \theta = 22°$

6. A tightrope walker is balanced on a wire strung between two buildings 100 feet apart. The wire dips at angles of 5° and 10° at the support points, as shown. What are the lengths of the segments of wire between where he is standing and each building? (Hint: Use law of sines.)

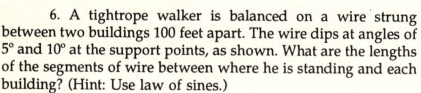

Angle at point B $= 180° - 5° - 10° = 165°$

Segment AB $= \sin 10° \left(\dfrac{100\ ft}{\sin 165°}\right) = 67\ ft$

Segment BC $= \sin 5° \left(\dfrac{100\ ft}{\sin 165°}\right) = 34\ ft$

7. A tugboat is moving a raft of logs by pulling on a tow rope with a force of 1600 lb at an angle of θ with respect to east. This force has a component along the N-S direction and a component along the E-W direction. Determine the size (in pounds) of each component and state whether its direction is N, S, E, or W, for the following values of θ.

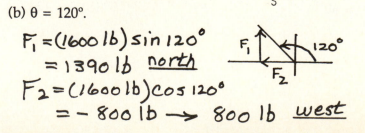

(a) $\theta = 60°$
$F_1 = (1600\ lb)\sin 60°$
$= 1390\ lb$ _north_
$F_2 = (1600\ lb)\cos 60°$
$= 800\ lb$ _east_

(b) $\theta = 120°$.
$F_1 = (1600\ lb)\sin 120°$
$= 1390\ lb$ _north_
$F_2 = (1600\ lb)\cos 120°$
$= -800\ lb \rightarrow 800\ lb$ _west_

Third Round Posttest — Optimum test time: 30 minutes or less

The following questions are word problems to test your ability to apply basic geometry and trigonometry to physical situations. Have scratch paper and a calculator ready before you begin. Put answers and required sketches in the right hand column. Then time yourself and check your responses against those given after the last question.

STARTING TIME _____   ANSWERS

1. A roof truss structure has two reinforcing members attached at 90° and 25° with respect to the other pieces as shown. What are the values of the indicated angles α, β, γ, and δ?

α = _____
β = _____
γ = _____
δ = _____

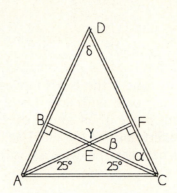

2. Refer to the roof truss structure described above. How can you argue that triangles ABE and AFD are similar? If AD = 20.0 ft, AF = 15.3 ft, and AB = 7.2 ft, what is the length AE?

_____

_____

3. A cube of metal which measures 1.0 inch on each edge has a circular hole of diameter 0.50 inch drilled through it perpendicular to a face. Draw a sketch as it appears in perspective. What is the volume of the metal which remains?

_____

4. A weight is suspended from the top of a triangular framework consisting of two 12.0 m long struts, each of which makes an angle of 70° with the horizontal. (a) How high from the ground is the point of attachment of the weight (apex of the triangle)? (b) How far apart on the ground are the feet of the struts?

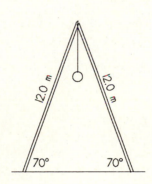

(a) _____

(b) _____

5. A steel sphere 2.4 cm in diameter rests in a V-shaped groove in a machine tool holder. The sides of the groove are symmetric with respect to the vertical and form a right angle. The sphere presses radially outward on the sides of the groove at each of the points of contact with a force F = 0.28 N (arrows). (a) What is the total of the downward components of force exerted by the sphere? (b) If steel weighs 7.9×10^{-2} N/cm³, what is the total weight of the sphere?

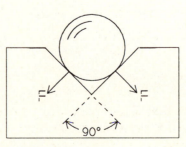

(a) _____

(b) _____

6. A barrel is rolled up a 10.0 ft long plank onto the bed of a truck. The truck bed is 5.0 ft above the ground. (a) What is the horizontal distance x from the lower end of the plank to the truck bed? (b) What is the angle θ which the plank makes with the ground?

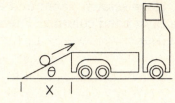

(a) _____

(b) _____

7. Between New Year's Day and June 1 the earth moves 150° in its orbit around the sun. (Assume a circular path with a radius of 1.5×10^8 km.) (a) How many km does the earth move in its course during this time? (b) Without using a calculator determine cos 150°. (c) Use the law of cosines to find the straight line distance between the earth's positions on these two dates.

(a) _____

(b) _____

(c) _____

ENDING TIME_____

ANSWERS:

1. α = 40°
   β = 50°
   γ = 130°
   δ = 50°

2. Angles are equal (40°, 50°, 90°).

   9.4 ft

3.

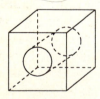

   0.80 in³

4. (a) 11.3 m

   (b) 8.2 m

5. (a) 0.57 N

   (b) 0.57 N

6. (a) 8.6 ft

   (b) 30°

7. (a) 3.9×10^8 km

   (b) $-\sqrt{3}/2$

   (c) 2.9×10^8 km

ESSAY: We Live in a Three-Dimensional World — Don't We?

We live and move about in a world which has three dimensions: up–down, forward–back, side-to-side. Physicists like to call this "three-space." This is a pervasive idea—so much so that it is considered to be one of the striking advances in both mathematics and physics when Rènè Descartes learned how to utilize the power of algebra to deal with the geometry of such space. According to legend, while soldiering in the Bavarian army, Descartes came upon this idea as he lay in bed watching a fly move about his room. The position of the fly at any moment could be described by three numbers—the perpendicular distances to the ceiling and to each of two walls. The path of the fly translates into a table of three-number combinations, or better yet, a mathematical function of x, y, and z. Even the shape of the fly, in principle, has an algebraic counterpart.

This "analytical geometry" of Descartes leads to much that is familiar—and useful—in the work of science. A point P in three-space is described by three numbers, but they need not be only x, y, and z (as shown in the drawing on the left below). The three numbers can be two angles (θ, ϕ) and the radial distance to the origin r, as shown in the right-hand drawing. Other combinations of three numbers are possible as well. And the fundamental facts about the world of three-space can be stated in each system. For instance, the shortest distance from the origin to P is r, but it is just as correctly written $\sqrt{x^2 + y^2 + z^2}$.

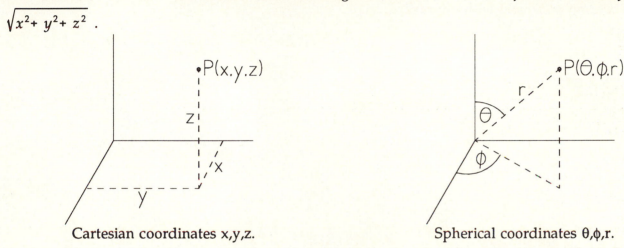

Cartesian coordinates x,y,z.   Spherical coordinates θ, ϕ, r.

Unfortunately, the images we make of three-dimensions are conveniently made only in a two-dimensional translation: the images cast on the retina of the eye, or drawings on a flat sheet of paper. But from our earliest experience we learn reasonably well to interpret these imperfect representations of the "real" world. On the other hand, it is fortunate that a lot of important phenomena are restricted to just two of the three space dimensions—and consequently require just two variables. In a sense, we stand "outside" 3-space and view these phenomena from the side; for example, the trajectory of baseball in flight is accurately portrayed on the page of a book. By extension, it is not too hard to use our imagination and the mathematics of 3-space to deal with a more complicated situation—perhaps a baseball in a cross wind.

There is insight to be gained by imagining creatures living in a two dimensional-world—maybe worm-like beings confined to the surface of a sphere with no concept of up and down. Viewing them from the vantage point of our 3-space we quickly recognize that these creatures have an unusual geometry. For instance the sum of the angles in their triangles do not have to total 180°. (The drawing on the next page shows a 2-space triangle with three right angles. It is a good thing the ancients could

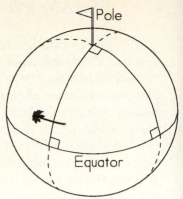

not measure triangles on the earth's surface more accurately—they may not have discovered the Pythagorean theorem!) The important point is that when dealing with different numbers of dimensions one has to be careful about the geometry. Fortunately the analytical geometry of Descartes makes us equal to the task. Thus creatures living in 4-space would use four coordinate numbers to describe a point in space. They would have no trouble viewing our world as a projection on three of their four dimensions.

Actually physics, as a matter of course, makes use of this idea of using four numbers to describe the way the world behaves. Physical events are not only specified by *where*, but *when*. Many of the equations you will see in physics contain not only position coordinates x, y, or z, but the time "coordinate" t. This is not to say that time and place are equivalent. Obviously, some physical happenings can be put in reverse so that objects retrace their paths back to their starting points, but never to their starting times. But no matter—the concept of using three distance measurements and a time measurement as the coordinates of a 4-space long ago proved its value in dealing creatively with physics.

Round IV — Physics of a Higher Order: Power Functions

Much of physics can be encapsulated into relatively simple mathematical equations. In this book an entire round was devoted to linear relationships, equations in which variables occur only to the first power. Algebraic expressions involving other powers also arise in beginning physics. This round emphasizes how to work with the most important power functions, especially quadratic equations. Begin by using this pretest to assess your skills in dealing with non-linear algebraic equations and formulas.

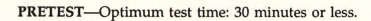

PRETEST—Optimum test time: 30 minutes or less.

Keep track of your time. Feel free to use a calculator when needed. Answers are given following the last question.

STARTING TIME_____   ANSWERS

1. Consider the following power functions:

$$y = ax^2 \qquad y = a/x \qquad y = a/x^2 \qquad y = a\sqrt{x}$$

    (a) On the graph sketch a typical curve representing each of these functions.

    (b) If $y = 2$ when $x = 3$, what is y when $x = 12$?

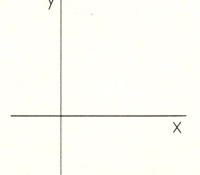

$y = $ _____

$y = $ _____

$y = $ _____

$y = $ _____

2. Solve the following equation and check your results by substituting each solution back into the original expression:

$$140 = 9x^2 + 12x .$$

$x = $ _____

3. Solve the following pair of simultaneous equations for x and y:

$$x = 5 - y \qquad and \qquad y = 2x^2 + 5x - 15 .$$

$(x,y)= $ _____

$(x,y)= $ _____

4. The product of two consecutive positive integers is 182. Find the integers.

_____

5. The width of a rectangle is 10 meters less than its length. If the area of the rectangle is 231 m^2, what are its dimensions?

_____

6. The pressure p of a certain sample of gas is found to vary inversely as its volume V. If the gas occupies 1.0 liter at atmospheric pressure, how much volume does it occupy when the pressure is doubled?

_____

7. Measurable quantities which are used in the following equations are length (x's), time (t's), and speed (v's). By checking for consistency in the dimensions of the terms in each equation, decide which of them cannot be correct:

(a) OK__NotOK___

(b) OK__NotOK___

(c) OK__NotOK___

(d) OK__NotOK___

(a) $x = \frac{1}{2}\left(\frac{v_o}{t}\right)$

(b) $v = \frac{x_2 - x_1}{t}$

(c) $v = \frac{2\pi x_r}{t_p}$

(d) $t = \frac{t_o}{\sqrt{v_L^2 - v^2}}$

ENDING TIME_____

ANSWERS:

1. (a)

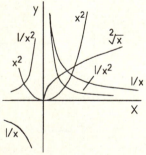

(b) 1/8, 8, 32, 1

2. 10/3, −4/3

3. (x,y) = −5,10

   (x,y) = 2,3

4. 13, 14

5. 21 m, 11 m

6. 0.5 liter

7. (a) Not OK

   (b) OK

   (c) OK

   (d) Not OK

Review 12 — Quadratic Expressions

A step up in complexity from linear equations are quadratic equations, "power functions" in which terms occur containing the independent variable raised to the second power. Expressions of this type are of especial importance in mechanics, which is the core subject of physics.

THE EQUATION AND ITS GRAPH

General expression. The most general form of a quadratic equation can be expressed as follows:

$$y = Ax^2 + Bx + C .$$

A, B, and C are constants, not depending on x, which determine in detail just how y will change as x takes on different values.

Squared proportion. In the simplest cases, both B and C equal zero. When the resulting equation

$$y = Ax^2$$

is plotted on a graph a symmetrical bowl-shaped curve called a *parabola* results. (For a negative value of A, the bowl is upside-down.) This equation represents a type of proportion; not a linear proportion, but a squared (or quadratic) proportion. But as with linear proportions we can often use ratios to find answers to problems, as in this example:

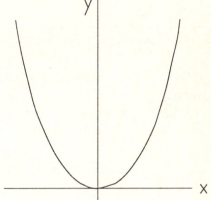

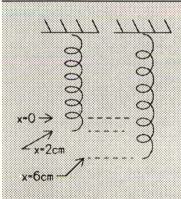

The energy stored in a simple helical spring depends on the amount x which the spring is stretched or compressed from its normal length. The relationship is

$$\text{Energy stored} = \tfrac{1}{2}kx^2$$

where the value of k, called the spring constant, depends on how the spring is constructed. If 10 joules (J) of energy is stored when a certain spring is extended 2.0 cm, how much energy is stored when the stretch is increased to 6.0 cm?

DISCUSSION: The stored energy is proportional to the square of the stretch. Therefore the ratio of the energies is equal to the ratio of squared stretches, as follows:

$$\frac{\text{Energy}(x=6\text{ cm})}{\text{Energy}(x=2\text{ cm})} = \frac{(6.0\text{ cm})^2}{(2.0\text{ cm})^2} .$$

Thus setting Energy(x=2 cm) = 10 J and rearranging, we have

$$\text{Energy}(x=6\text{ cm}) = (10\text{ J})(36/4) = 90 \text{ joules.}$$

Other parabolas. When B and C are not both zero, the graphs of the quadratic equation are also parabolic but may be shifted up, down, or sideways, or flipped upside down (but never rotated), depending on the values of A, B, and C. Three cases are illustrated here.

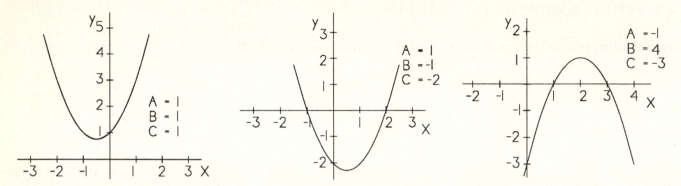

The approximate width of the parabola and its location with respect to the axes may not be readily apparent, given values of A, B, and C, but a couple of facts may occasionally be useful for a quick graphical interpretation of a result: (a) negative values of A give rise to "upside-down" parabolas, and (b) C is the y-intercept (the value of y where the curve crosses the y-axis).

SOLVING QUADRATIC EQUATIONS

Standard form of the equation and roots. Except for the point at the very bottom (or top) of the curves where each of the parabolas shown above turn around, there are *two values* of x corresponding to every possible value of y. Of particular interest are the two values of x where the curve crosses the x-axis. These are the two solutions or "roots" of an expression which has the following form:

$$0 = Ax^2 + Bx + C .$$

Sometimes the algebraic statement of a physics problem gives us this special case of a quadratic equation directly. However, in general, a non-zero value of y is one of the given quantities and we must rearrange the equation into the above "standard form" in order to find the values of x which satisfy the problem. Stated briefly:

> *Substitute the known value of y into the quadratic equation which represents the physics. To find the two roots x, it is generally useful to algebraically rearrange the equation into "standard form"* $0 = Ax^2+Bx+C$.

Finding a pair of roots as the solution of a physical problem is illustrated by the following example. (This particular case is simple enough mathematically that it isn't necessary to write the quadratic equation in standard form.)

> The energy of a spring which is stretched or squeezed an amount x (in cm) from its normal length is given by
>
> $$E = (4 \text{ joules/cm}^2)x^2 .$$
>
> At what values of x is the energy 16 joules?
>
> DISCUSSION: (Next page.)

DISCUSSION: Substitute for E to give 16 J = (4 J/cm²)x². Rearranging we have

$$x^2 = 4 \text{ cm}^2.$$

Taking the square root on both sides of the equation yields the two roots

$$x = +2 \text{ cm (stretch)}$$

and

$$x = -2 \text{ cm (squeeze)}.$$

The quadratic formula. In the last example, the constant B was zero; consequently finding the solutions required no special technique except finding an ordinary square root. However, in general, roots of the equation $0 = Ax^2+Bx+C$ can always be obtained by applying the following "quadratic formula:"

$$x = \frac{-B \pm\sqrt{B^2 - 4AC}}{2A}$$

This formula is so general and useful as to be worth memorizing. In the following example illustrating the use of the quadratic formula, units have been left out in order to clarify the mathematical manipulation. Problems which involve units are included in the accompanying drill and are especially emphasized in the next review.

In a certain problem y and x are related by the expression y-1 = x(4x+2). What are the values of x corresponding to y = 3?

DISCUSSION: The expression can be rewritten as

$$y = 4x^2 + 2x + 1.$$

Substituting y=3, the equation can be rearranged into the standard form

$$0 = 4x^2 + 2x - 2.$$

Identifying A=4, B=2, and C=-2, the quadratic formula gives for the two roots

$$x = \frac{-2 \pm\sqrt{4-(4)(4)(-2)}}{(2)(4)}$$

$$= \tfrac{1}{2} \text{ and } -1.$$

Two roots are always to be expected, although one of the solutions sometimes is rejected because it is not meaningful physically. Occasionally both roots are identical; graphically such a solution corresponds to a point located at the turning point (bottom of the "bowl") of the parabola.

SIMULTANEOUS EQUATIONS

Conditions for a solution. Review 7 considered problems in which more than one unknown quantity could be found using a set of several linear equations. To do this the following requirement must be met:

There must be at least as many independent simultaneous equations as there are unknown variables.

This rule also applies when a set of simultaneous algebraic equations are not linear, although the algebra needed in some cases may be somewhat complicated. Fortunately, almost all instances you are liable to encounter in introductory physics involving squared variables can be handled in a straightforward way using a "substitution of equations" approach.

Simultaneous quadratic and linear equations. When one of the equations is quadratic and others are linear it is relatively easy to eliminate all but one of the variables from the higher power equation. It can then be solved using the usual methods for solving quadratic equations. This is illustrated by the following strictly algebraic example.

Find the values of the two unknowns x and y which simultaneously satisfy the following two equations:

$$y = x^2 + 3x - 6$$
$$y = x + 2$$

DISCUSSION. The value of y from the second equation can be substituted into the first equation to give

$$x + 2 = x^2 + 3x - 6 ,$$

or

$$0 = x^2 + 2x - 8.$$

The two roots of this equation can be found using the quadratic formula; these are $x = 2$ and -4. Next, the corresponding values of y are found by substituting each of these values of x into one of the original equations. This gives us two solution pairs, which are

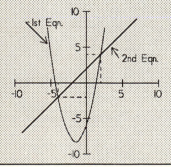

$$(x,y) = (2,4) \text{ and } (-4,-2).$$

(These results may be checked by substituting back into each of the original equations.) Each solution pair corresponds to a point of intersection of the curves which represent the two simultaneous equations. In this example there are two solution pairs, hence two points of intersection, as shown in the graph.

Skill Drill 12

The exercises in this drill emphasize the algebra of quadratic equations, with a variety of applications to word problems. Quadratic equations will be revisited in the next review, which deals with the intelligent use of formulas in physics.

1. Review of major points. Consider the quadratic expression

$$y = ax^2 + bx + c .$$

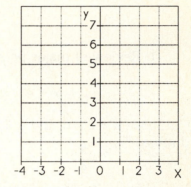

(a) Suppose $b = c = 0$. If $y = 4$ for a certain value of x, what is y when x is doubled?

(b) Again assuming $b = c = 0$ plot y versus x on the accompanying graph, for $a = 0.5$. From your plot find x when $y = 5$.

(c) Now take $a = 0.5$, $b = 1$, and $c = 2$. Before plotting the equation, determine the value of y at which the curve will cross the vertical axis. Now plot the equation on the same graph which was used in question (b).

(d) Again assume the values of the constants given in part (c)
Use the "quadratic formula" to find x when $y = 14$.

(e) For the values of the constants given in part (c) find x when the simultaneous equation $y = 2x + 4$ is also satisfied.

2. Solve each of the following equations and check your results by substituting back into the original expression:

(a) $x^2 + 3x + 2 = 0$

(b) $x(5x - 1) = (x + 1)(5x + 2)$

(c) $x^2 - 6x + 9 = 0$

(d) $16x^2 - 48x = 0$

Skill Drill 12 — SOLUTIONS AND ANSWERS

1. Review of major points. Consider the quadratic expression

$$y = ax^2 + bx + c .$$

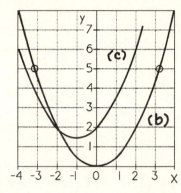

(a) Suppose b = c = 0. If y = 4 for a certain value of x, what is y when x is doubled?

$$\frac{y}{4} = \left(\frac{2x}{x}\right)^2 \longrightarrow \quad y = 16$$

(b) Again assuming b = c = 0 plot y versus x on the accompanying graph, for a = 0.5 . From your plot find x when y = 5.

calculate y at x = ± 0, 1, 2, 3, 4 and draw curve.
 From curve x = ± 3.2 for y = 5 (circled).

(c) Now take a = 0.5, b = 1, and c = 2. Before plotting the equation, determine the value of y at which the curve will cross the vertical axis. Now plot the equation on the same graph which was used in question (b).

Curve crosses y-axis at y = c = 2.
Draw curve using values of y as in table at right.

x	y
2	6
1	3 1/2
-1	1 1/2
-2	3 1/2
-4	6

(d) Again assume the values of the constants given in part (c). Use the "quadratic formula" to find x when y = 14.

$$14 = \tfrac{1}{2}x^2 + x + 2$$
$$0 = \tfrac{1}{2}x^2 + x - 12$$
$$x = \frac{-1 \pm \sqrt{1 - 4(-12)/2}}{1}$$
$$= -1 \pm \sqrt{25} = 4, -6$$

(e) For the values of the constants given in part (c) find x when the simultaneous equation y = 2x + 4 is also satisfied.

$$y = \tfrac{1}{2}x^2 + x + 2$$
$$y = 2x + 4$$

subtracting yields

$$0 = \tfrac{1}{2}x^2 - x - 2$$
$$x = 1 \pm \sqrt{1 + 4(\tfrac{1}{2})2}$$
$$= 1 \pm \sqrt{5} = 3.2, -1.2$$

2. Solve each of the following equations and check your results by substituting back into the original expression:

(a) $x^2 + 3x + 2 = 0$

$$x = -\frac{3 \pm \sqrt{9 - 4(2)}}{2}$$
$$= -3/2 \pm 1/2 = -1, -2$$

Checking:
 For x = -1, lhs = 1 - 3 + 2 = 0
 For x = -2, lhs = 4 - 6 + 2 = 0

(b) x(5x − 1) = (x + 1)(5x + 2)

$$\longrightarrow 5x^2 - x = 5x^2 + 5x + 2x + 2$$
or $-8x = 2$ which is linear.
$$\longrightarrow x = -1/4 \text{ (single root)}$$
Checking:
$$-\tfrac{1}{4}\left(-\tfrac{5}{4} - 1\right) = \left(\tfrac{3}{4}\right)\left(-\tfrac{5}{4} + 2\right)$$
$$9/16 = 9/16$$

(c) $x^2 - 6x + 9 = 0$

$$x = \frac{6 \pm \sqrt{36 - 4(9)}}{2}$$
$$= 3, 3 \text{ (identical roots)}$$
Checking:
 with x = 3, lhs = 9 - 18 + 9 = 0

(d) $16x^2 - 48x = 0$

Factor out 16x.
$$\longrightarrow 16x(x - 3) = 0$$
Roots are x = 0, 3.
Checking:
 For x = 3, lhs = 144 - 144 = 0

(e) $8x^2 + 26x = -21$ (f) $x^2 - 1 = (5/6)x$

3. Solve each of the following pairs of simultaneous equations:

(a) $5x^2 + 15x - 6 = y$ (b) $y = x^2$
    $-y = 6x + 10$. $y = 45 - 12x$.

4. The product of two consecutive even numbers is 224. What are the numbers?

5. The sum of two numbers is 9, and the difference of their squares is 45. What are the numbers?

6. The combined areas of two squares is 424 in^2. If the side of one square is 8 inches greater than the side of the other square, what are the dimensions of both figures?

7. Ralph made a trip of 200 miles at an average rate of speed v. Had he decreased his average speed by 10 mi/hr the same trip would have taken 1 hour more. Find v.

(e) $8x^2 + 26x = -21$

$$\to 8x^2 + 26x + 21 = 0$$

$$x = \frac{-26 \pm \sqrt{676 - 672}}{16}$$

$$= -13/8 \pm 1/8 = -7/4, -3/2$$

checking:

For $x = -7/4$: $24\frac{1}{2} - 45\frac{1}{2} + 21 = 0$

For $x = -3/2$: $18 - 39 + 21 = 0$

(f) $x^2 - 1 = (5/6)x$

$$\to x^2 - (5/6)x - 1 = 0$$

$$x = \frac{5/6 \pm \sqrt{25/36 + 4}}{2}$$

$$= 5/6 \pm \sqrt{169/36} = 3/2, -2/3$$

Checking: For $x = 3/2$:

$9/4 - 1 = 15/12 \to 5/4 = 5/4$

For $x = -2/3$: $4/9 - 1 = -10/18$

$$\to -5/9 = -5/9$$

3. Solve each of the following pairs of simultaneous equations:

(a) $5x^2 + 15x - 6 = y$

$-y = 6x + 10$.

Substitute 2nd eqn into 1st eqn:

$-6x - 10 = 5x^2 + 15x - 6$

$\to 0 = 5x^2 + 21x + 4$

Quadratic formula:

$$x = (-21 \pm \sqrt{441 - 80})/10$$

$$= (-21 \pm 19)/10 = -1/5, -4$$

From 2nd eqn:

$y = -6(-1/5, -4) - 10 = -44/5, 14$

(b) $y = x^2$

$y = 45 - 12x$.

$x^2 = 45 - 12x$

$\to x^2 + 12x - 45 = 0$

Quadratic formula:

$$x = (-12 \pm \sqrt{144 + 180})/2$$

$$= -6 \pm 9 = 3, -15$$

using 1st eqn:

$(x, y) = (3, 9)$ and $(-15, 225)$

4. The product of two consecutive even numbers is 224. What are the numbers?

Let one number $= x$;   The other is $x + 2$.

$$x(x + 2) = 224 \to x^2 + 2x - 224 = 0$$

Quadratic formula:

$$x = (-2 \pm \sqrt{4 + 896})/2 = (-2 \pm 30)/2 = 14, -16$$

Thus either $x = 14$, $x + 2 = 16$ or $x = -16$, $x + 2 = -14$

5. The sum of two numbers is 9, and the difference of their squares is 45. What are the numbers?

Call the two numbers x and y

$x + y = 9$

$x^2 - y^2 = 25$

substituting eqn 1 into eqn 2:

$x^2 - (81 + x^2 - 18x) = 45$

$18x = 126$

This is linear with one solution pair:

$x = 7$

$y = 2$

6. The combined areas of two squares is 424 in^2. If the side of one square is 8 inches greater than the side of the other square, what are the dimensions of both figures?

$x^2 + (x + 8 \text{ in})^2 = 424 \text{ in}^2$

Drop units temporarily:

$2x^2 + 16x + 64 = 424$

$\to 2x^2 + 16x - 360 = 0$

Quadratic formula

$$x = \frac{-16 \text{ in} \pm \sqrt{256 \text{ in}^2 + 2880 \text{ in}^2}}{4}$$

$= -4 \text{ in} \pm 14 \text{ in} = 10 \text{ in}, -18 \text{ in}$

Negative root not meaningful.

Square A: 10 in × 10 in

Square B: 18 in × 18 in

7. Ralph made a trip of 200 miles at an average rate of speed v. Had he decreased his average speed by 10 mi/hr the same trip would have taken 1 hour more. Find v.

$200 \text{ mi} = vt$

also

$200 \text{ mi} = (v - 10 \text{ mph})(t - 1 \text{ hr})$

Eliminate t from pair. Also supress units until last steps.

$0 = (v - 10)(200/v - 1)$

$0 = (v - 10)(200 - v)$

$0 = v^2 - 10v - 2000$

$$v = \frac{10 \text{ mph} \pm \sqrt{100 \text{ mph}^2 + 800 \text{ mph}^2}}{2}$$

Positive root: $v = 50 \text{ mph}$

8. Farmer Smith tethers her goat on a rope attached to a post so that the goat can graze over a circular area of meadow. If the length of the tether rope is increased by 100 feet the goat can graze over a circle with 9 times the area. What is the length of the original rope?

9. A certain ravine has a parabolic cross-section, as illustrated. If the ravine is 300 feet across at a height 100 feet above the bottom, what is the width at the top of the ravine 200 feet above the bottom?

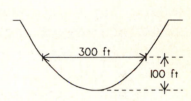

10. The so-called "potential" energy E of a system consisting of a spring with a weight W suspended from it is given by

$$E = \tfrac{1}{2}kx^2 - Wx$$

where x is the difference between the length of the spring and its length before the weight is attached. If the spring constant k = 10 lb/ft and W = 30 lb, what values of x correspond to an energy E = 2.5 ft-lb?

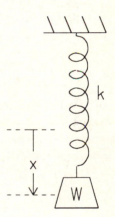

11. A billiard ball collides head-on with a stationary marble. Using laws of mechanics it is possible to write equations which relate the speeds of the billiard ball and marble after the collision (v_B and v_M, respectively) to the incoming speed of the billiard ball. Suppose for a certain case the equations are as follows:

$$10\ v_B + v_M = 150 \text{ cm/s} \qquad \textit{and} \qquad 10\ v_B^2 + v_M^2 = 10\ (15 \text{ cm/s})^2\ .$$

Solve this pair of simultaneous equations to determine v_M.

8. Farmer Smith tethers her goat on a rope attached to a post so that the goat can graze over a circular area of meadow. If the length of the tether rope is increased by 100 feet the goat can graze over a circle with 9 times the area. What is the length of the original rope?

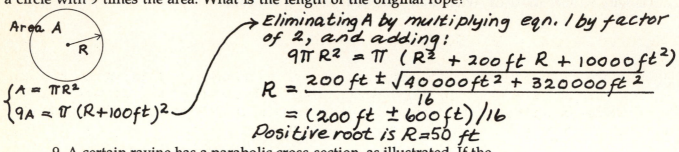

Area A

$\begin{cases} A = \pi R^2 \\ 9A = \pi (R+100ft)^2 \end{cases}$

→ *Eliminating A by multiplying eqn. 1 by factor of 2, and adding:*

$9\pi R^2 = \pi (R^2 + 200 ft R + 10000 ft^2)$

$R = \dfrac{200 ft \pm \sqrt{40000 ft^2 + 320000 ft^2}}{16}$

$= (200 ft \pm 600 ft)/16$

Positive root is R = 50 ft

9. A certain ravine has a parabolic cross-section, as illustrated. If the ravine is 300 feet across at a height 100 feet above the bottom, what is the width at the top of the ravine 200 feet above the bottom?

Put origin at bottom of ravine, so sidewalls are described by

$y = A x^2$

using ratios $y'/y = (x'/x)^2$ *where primes refer to top of ravine.*

$x' = x\sqrt{y'/y} = 150 ft\sqrt{200/100} = 212 ft$

Width at top = 2x' = 424 ft

10. The so-called "potential" energy E of a system consisting of a spring with a weight W suspended from it is given by

$$E = \tfrac{1}{2}kx^2 - Wx$$

where x is the difference between the length of the spring and its length before the weight is attached. If the spring constant k = 10 lb/ft and W = 30 lb, what values of x correspond to an energy E = 2.5 ft-lb?

Quadratic eqn. in standard form:

$0 = k x^2 - 2Wx - 2E$

$X = (2W \pm \sqrt{4W^2 + 8k E})/2k$

$= \dfrac{60 lb \pm \sqrt{3600 lb^2 + 200 lb^2}}{20 lb/ft} = +6.1 ft, -0.082 ft$

11. A billiard ball collides head-on with a stationary marble. Using laws of mechanics it is possible to write equations which relate the speeds of the billiard ball and marble after the collision (v_B and v_M, respectively) to the incoming speed of the billiard ball. Suppose for a certain case the equations are as follows:

$$10 v_B + v_M = 150 \text{ cm/s} \qquad and \qquad 10 v_B^2 + v_M^2 = 10 (15 \text{ cm/s})^2 .$$

Solve this pair of simultaneous equations to determine v_M.

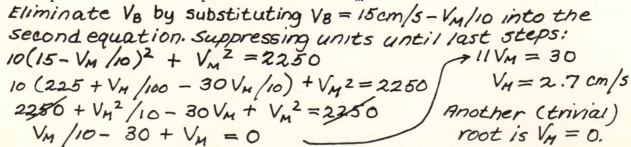

Eliminate V_B by substituting $V_B = 15 cm/s - V_M/10$ into the second equation. Suppressing units until last steps:

$10(15 - V_M/10)^2 + V_M^2 = 2250$

$10 (225 + V_M^2/100 - 30 V_M/10) + V_M^2 = 2250$

$2250 + V_M^2/10 - 30 V_M + V_M^2 = 2250$

$V_M/10 - 30 + V_M = 0$

→ $11 V_M = 30$

$V_M = 2.7 cm/s$

Another (trivial) root is $V_M = 0$.

Review 13 — Problems, Formulas, and Physics

The art of successful problem solving in physics cannot be mastered simply by memorizing lists of useful formulas. Nevertheless, formulas are required to find most numerical answers. Using an example problem whose solution involves a quadratic equation, this review shows how formulas are used to solve problems in a discriminating and error-free way.

A QUADRATIC EQUATION FROM PHYSICS

Uniform acceleration. Early in your physics course you will encounter a quadratic expression which describes the motion of objects whose speed along a straight path changes at a constant rate. Such an object is said to be undergoing *uniform acceleration*. An example of such behavior is the motion of a weight which is freely falling under the influence of gravity.

The equation and an example problem. The following quadratic equation describes how the position y of an object undergoing uniform acceleration depends on the time t:

$$y = \tfrac{1}{2}at^2 + v_o t + y_o$$

The constants v_o and y_o are, respectively, the speed and position of the object when the time t = 0 (when, for instance, a timing clock is started). The constant a, called *acceleration*, is a measure of the rate at which speed changes. A positive value of acceleration means that speed increases during motion in the +y direction, whereas a negative value means that speed decreases during such motion. Our concern at this point, however, is less on interpretation than on the successful use of the expression.

The following is a typical problem having to do with uniform acceleration:

> A ball is thrown straight downwards from the top of a 30 m high building with an initial speed of 5.0 m/s. Knowing that such a freely falling body has a constant acceleration a = 9.8 meters per second per second (9.8 m/s^2), determine how many seconds it takes for the ball to reach the ground.

Broadly speaking, the solution to a problem like this depends on correctly identifying both the given and unknown quantities and relating them through a valid equation (or equations) of physics. The answer is then obtained by solving and inserting numerical values in these equations. How this process can be carried out efficiently and correctly will be illustrated by reference to this example problem in the next section.

INTELLIGENT USE OF FORMULAS

Several features commonly characterize the successful application of equations in the solution of physics problems. These characteristics are discussed below.

(1) Making and labeling a sketch. In many cases even the most rudimentary drawing will help in visualizing and understanding a problem. This is often the best first step in organizing the information to be used in the solution.

As an example, the problem given above is illustrated by a sketch given below. Notice particularly that

> • *factual information is very conveniently summarized in a sketch.*

In this drawing, the symbol for the initial speed v_o is used as a label on a down-pointing arrow at the top of the building. Also, the main objective of the problem is summarized by "t=?" written at the bottom of the building. Notice also that in this case, as in other problems in which directions in space are important,

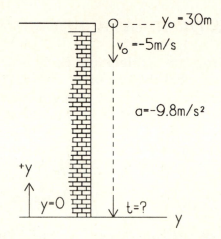

> • *the choice of coordinate axes is explicitly indicated in the drawing.*

In the example diagrammed here, the y-axis has been chosen to point upwards; an arrow labelled "+y" indicates this. Likewise, the choice of the origin (y = 0) at the base of the building has been clearly indicated. It should be understood that the choice of coordinate directions and origin are largely arbitrary. Certainly the final answer to a problem should not depend on an arbitrary choice, although the numbers which must be used later in the solution may reflect that decision.

(2) Choosing the equation(s). With an understanding of the problem well in hand you are ready to ask yourself some crucial questions:

> • *What is the given information and what information is to be determined?*

> • *Can I think of an equation which relates these? Or, do I know a relationship which gives an intermediate result which can be used in another equation to get an answer?*

> • *Is there any useful information, not explicitly stated in the problem, which can be tacitly assumed?*

Although simple rote memorization of formulas is discouraged, with experience the most important equations inevitably will be remembered. Often the problem, while not identical to any of them, will remind you of problems you have done in the past. In any event in choosing a relationship

> • *one should always be guided by an understanding of the underlying physics and of the limits of applicability of the equation.*

For instance, the equations which apply to the above problem must have to do with *uniformly-accelerated* motion since this describes freely falling objects; such equations cannot be used if acceleration is non-uniform. The solution of the problem continues as follows:

> The given quantities are the initial position y_o and initial speed v_o, as well as the position y at the end of the falling motion. The acceleration a is also a known quantity. An equation which relates all of these to the unknown time t at the end of the fall is the quadratic expression
>
> $$y = \tfrac{1}{2}at^2 + v_o t + y_o .$$

(3) Checking on the equation. Have you remembered the equation correctly?

The equation should appear to make sense, or at least it should not appear to be nonsense.

For instance, in the example, you would be amiss to forget the term involving v_o; the distance an object has moved after some time has elapsed surely depends on how fast it starts off. Moreover

- *the equation should be dimensionally correct; every term should have the same units.*

In the example the term $\frac{1}{2}at^2$ has the dimensions of distance, as expected: (meter/second^2)×(second)^2 = meters. (Checking dimensions does not, of course, allow us to obtain the dimensionless factor $\frac{1}{2}$.)

(4) Calculating the answer.

In most cases it is more efficient to do the necessary algebra before substituting in numbers.

The reason for this hint is that less writing is required, cancellation of terms is easily recognized, and simple arithmetic can be performed while the expression is least cumbersome. Finally

- *numerical quantities should be substituted <u>with units included</u>.*

The numbers which are substituted must take into account the coordinate axes which you have elected to use. In the drawing the $+y$ axis points upwards whereas v_o points downwards. Consequently v_o has a negative numerical value. Likewise the acceleration a has a negative numerical value, since a weight rising in the $+y$ direction would be slowed down by gravity. The choice of origin sets $y = 0$. (Practice using other choices of axes is given in the Skill Drill.) In keeping with these pointers, the remaining step in the solution of the example problem is as follows:

In "standard form" the quadratic expression given on the previous page is

$$0 = \frac{1}{2}at_2 + v_ot + (y_o-y)$$

from which we can identify the coefficients $A = \frac{1}{2}a$, $B = v_o$, and $C = y_o-y$. The numerical values which are consistent with the choice of coordinates are $v_o = -5$ m/s, $y_o = 30$ m, and $y = 0$. The acceleration $a = -9.8$ m/s^2. Hence, using the quadratic formula for the roots of the equation, we have

$$t = \frac{-v_o \pm \sqrt{v_o^2 - 2a(y_o - y)}}{a}$$

$$= \frac{5.0\text{m/s} \pm \sqrt{(-5.0\text{m/s})^2 - 2(-9.8\text{m/s}^2)(30\text{m}-0)}}{-9.8\text{m/s}^2}$$

$$= 2.0 \text{ s, and } -3.0 \text{ s} .$$

The first of these two roots corresponds to the desired answer; the ball reaches the ground in 2.0 seconds.

Notice that using units along with numbers in the final calculation shown above provides a way of checking both the dimensions of the terms and the units in the answer, which should always be written down. Finally

> • *one should always ask: Does the answer seem reasonable?*

For instance, in this example it is likely that your experience with similar problems will lead you to realize that *seconds* is the right order of magnitude of the time for a body to fall the height of a building.

(5) Asking further questions. Try to engage physics problems with an inquisitive spirit, particularly in homework assignments. For instance, you might wonder whether there is some meaning to the negative root (-3.0 s) obtained in the example above. Although most of the time a graph is not needed, in this case a quickly drawn graph can be a guide to understanding, as discussed below.

GRAPHS AND INTERPRETATION

Graph of the falling body problem. The quadratic equation representing uniformly accelerated motion

$$y = \tfrac{1}{2}at^2 + v_o t + y_o$$

plots as a parabola on a graph of position y versus time t. The curve (taking the choice of axes as in the last section) is shown here.

Review 12 gave some useful tips which could be used for making an approximate sketch of this graph: the coefficient $\tfrac{1}{2}a$ is negative, so the parabola is an upside-down bowl; also the constant term y_o is the intercept. Since the body continues to fall after leaving the top of the building the top of the "bowl" must be to left of the t=0 position.

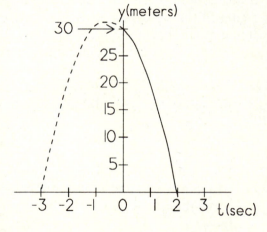

Interpretation. The original problem asked the question: when does the ball strike the ground? The answer, t = 2.0 s, corresponds to the point on the graph where the parabola passes through the horizontal (y = 0) axis.

The part of the parabola in the -t region of the graph is plotted as a dashed curve; the "mysterious" answer, t = -3.0 s, corresponds to the point where the dashed parabola passes through the horizontal axis. Hence we are led to the following interpretation of the negative root: a ball rising from the ground with an appropriate speed 3 seconds *before* we start our stopwatch (t = -3.0 s) would reach a maximum height, and then at t = 0 be falling with a speed v_o. Of course, it would again reach the ground at t = 2.0 s, as the curve illustrates.

Skill Drill 13

This drill asks you to give further thought to some of the hints for the intelligent use of formulas outlined in the previous review.

1. The following diagrams all apply to the problem of a freely falling ball used as an example in Review 13. In each case the choice of direction and origin of the coordinate axis is different. Otherwise the problem is identical to the example.

Give the values (with sign) of y_o, y (at base of building), v_o, and a which you would use in the quadratic expression given in the review to find the time the ball reaches the ground. (For numerical values, refer to the statement of the problem on the first page of Review 13.)

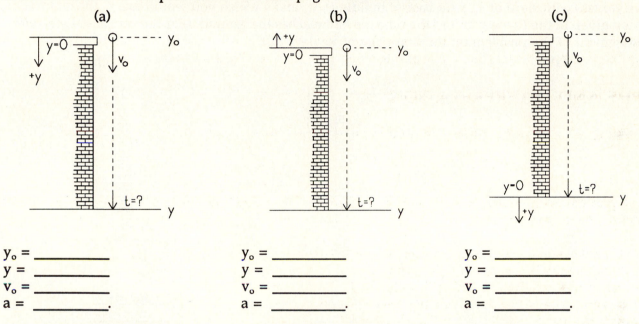

(a)   (b)   (c)

y_o = _____ y_o = _____ y_o = _____

y = _____ y = _____ y = _____

v_o = _____ v_o = _____ v_o = _____

a = _____ . a = _____ . a = _____ .

2. Verify that the choice of y_o, y, v_o, and a appropriate to Figure (a) above yields the same value of time ($t = 2.0$ s) which was found in the example problem in Review 13. Recall that the equation to use is

$$y = \tfrac{1}{2}at^2 + v_ot + y_o .$$

3. The following is an assortment of equations which might be encountered in connection with problems in several areas of physics. Some of them are correct and some are erroneous. Apart from subscripts (and primes) all of the equations involve position (or length) x, speed v, acceleration a, and time t. By checking the dimensions of the terms in these expressions, determine which are possibly correct and which are surely incorrect. (Remember: trig functions are dimensionless.)

Skill Drill 13 — SOLUTIONS AND ANSWERS

1. The following diagrams all apply to the problem of a freely falling ball used as an example in Review 13. In each case the choice of direction and origin of the coordinate axis is different. Otherwise the problem is identical to the example.

Give the values (with sign) of y_o, y (at base of building), v_o, and a which you would use in the quadratic expression given in the Review to find the time the ball reaches the ground. (For numerical values, refer to the statement of the problem on the first page of Review 13.)

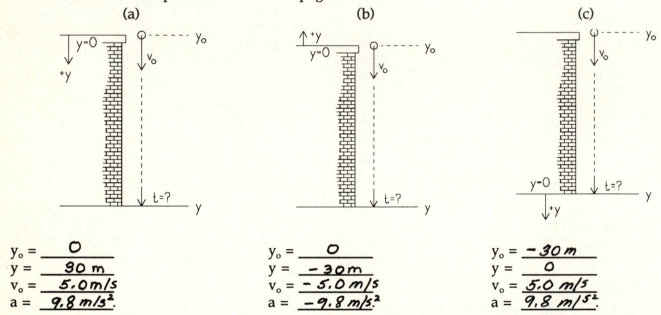

(a)

$y_o =$ _____ 0 _____
$y =$ _____ $30 \, m$ _____
$v_o =$ _____ $5.0 \, m/s$ _____
$a =$ _____ $9.8 \, m/s^2$ _____

(b)

$y_o =$ _____ 0 _____
$y =$ _____ $-30 m$ _____
$v_o =$ _____ $-5.0 \, m/s$ _____
$a =$ _____ $-9.8 \, m/s^2$ _____

(c)

$y_o =$ _____ $-30 \, m$ _____
$y =$ _____ 0 _____
$v_o =$ _____ $5.0 \, m/s$ _____
$a =$ _____ $9.8 \, m/s^2$ _____

2. Verify that the choice of y_o, y, v_o, and a appropriate to Figure (a) above yields the same value of time (t = 2.0 s) which was found in the example problem in Review 13. Recall that the equation to use is

$$y = \tfrac{1}{2}at^2 + v_o t + y_o \ .$$

$$\tfrac{1}{2}at^2 + v_o t + (y_o - y) = 0$$

$$\tfrac{1}{2}(9.8 \, m/s^2)t^2 + (5.0 \, m/s)t - 30m = 0$$

$$t = \frac{-5.0 \, m/s \pm \sqrt{25 \, m^2/s^2 - 4(4.9 \, m/s^2)(-30 \, m)}}{9.8 \, m/s^2}$$

$$= \frac{-5.0 \, m/s \pm 24.8 \, m/s}{9.8 \, m/s^2} = 2.0s, -3.0s$$

Taking positive root: $t = 2.0s$

3. The following is an assortment of equations which might be encountered in connection with problems in several areas of physics. Some of them are correct and some are erroneous. Apart from subscripts (and primes) all of the equations involve position (or length) x, speed v, acceleration a, and time t. By checking the dimensions of the terms in these expressions, determine which are possibly correct and which are surely incorrect. (Remember: trig functions are dimensionless.)

$x = \frac{1}{2}v_o/t$ OK__Not OK__ $\tan \theta = \dfrac{v^2}{a_g\, x_r}$ OK__Not OK__

$t = \sqrt{\dfrac{2(x-x_o)}{a}}$ OK__Not OK__

                                                                         $t_p = \dfrac{v-v_s}{x_w}$ OK__Not OK __

$v_{av} = \dfrac{v_1 + v_2}{2t}$ OK__Not OK__

                                                                         $v_R = \dfrac{\acute{v} + v}{1+\dfrac{\acute{v}v}{v_L^2}}$ OK__Not OK __

$x = \dfrac{v_o^2}{a_g}\ \sin2\theta$ OK__Not OK__

$t_p = 2\pi\sqrt{\dfrac{a_g}{x_p}}$ OK__Not OK__ $t_p = t_o\sqrt{\dfrac{1-v/v_0}{1+v/v_0}}$ OK__Not OK __

4. This problem is a variation on the example problem discussed in Review 13:

At the same moment (t=0) that the ball is thrown downwards from the top of the 30 meter high building with a speed of 5.0 m/s, an elevator on the side of the building starts upwards at a *constant* speed of 5.0 m/s. How long afterwards (time *t*) do the ball and elevator pass by each other? This can be found by solving the following pair of simultaneous equations for t:

$$y = \tfrac{1}{2}at^2 + v_{oB}t + y_{oB} \qquad and \qquad y = v_{oE}t + y_{oE}\ .$$

where subscripts B and E refer to the positions and speeds of the ball and elevator, respectively. y is the position at which ball and elevator pass each other.

Draw a diagram which illustrates this problem . Indicate on the drawing the origin and direction of the y-axis, as well as the values (with signs) of v_{oB}, y_{oB}, v_{oE}, and y_{oE}.

Designate length L, time T, speed L/T, acceleration L/T²

$x = \frac{1}{2}v_o t$ OK__ Not OK ✓

$$L \neq L/T^2$$

$t = \sqrt{\dfrac{2(x-x_o)}{a}}$ OK ✓ Not OK__

$$T = \sqrt{L \Big/ \frac{L}{T^2}}$$

$v_{av} = \dfrac{v_1 + v_2}{2t}$ OK__ Not OK ✓

$$L/T \neq \frac{L/T}{T}$$

$x = \dfrac{v_o^2}{a_g}\sin 2\theta$ OK ✓ Not OK__

$$L = (L^2/T^2)/(L/T^2)$$

$t_p = 2\pi\sqrt{\dfrac{a_g}{x_p}}$ OK__ Not OK ✓

$$T \neq \sqrt{(L/T^2)/L}$$

$\tan\theta = \dfrac{v^2}{a_g x_r}$ OK ✓ Not OK__

$$No\ dimensions = \frac{L^2/T^2}{(L/T^2)L}$$

$t_p = \dfrac{v - v_s}{x_w}$ OK__ Not OK ✓

$$T \neq \frac{L/T}{L}$$

$v_R = \dfrac{\acute{v} + v}{1 + \dfrac{v\acute{v}}{v_L^2}}$ OK ✓ Not OK__

$$L/T = \frac{L/T}{No\ dimens.}$$

$t_p = t_o\sqrt{\dfrac{1 - v/v_0}{1 + v/v_0}}$ OK ✓ Not OK__

$$T = T\sqrt{No\ dimens.}$$

4. This problem is a variation on the example problem discussed in Review 13:

At the same moment (t=0) that the ball is thrown downwards from the top of the 30 meter high building with a speed of 5.0 m/s, an elevator on the side of the building starts upwards at a *constant* speed of 5.0 m/s. How long afterwards (time *t*) do the ball and elevator pass by each other? This can be found by solving the following pair of simultaneous equations for t:

$$y = \tfrac{1}{2}at^2 + v_{oB}t + y_{oB} \qquad and \qquad y = v_{oE}t + y_{oE}\ .$$

where subscripts B and E refer to the positions and speeds of the ball and elevator, respectively. y is the position at which ball and elevator pass each other.

Draw a diagram which illustrates this problem . Indicate on the drawing the origin and direction of the y-axis, as well as the values (with signs) of v_{oB}, y_{oB}, v_{oE}, and y_{oE}.

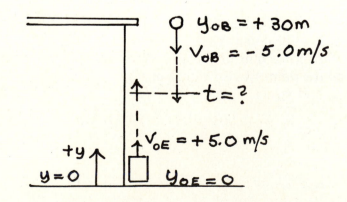

Review 14 — Other Power Laws

Some important physics is described by power functions other than linear and quadratic. Especially important are expressions involving negative powers—inverse functions—since such relationships often apply to phenomena in which the increase in one quantity is accompanied by a decrease in another quantity. This review deals with these and other power functions which come up in a course in introductory physics.

CUBICS AND QUARTICS

Power laws and ratios. The graphs of the power functions

$$y = Ax^3 \text{ and } y = Ax^4$$

are similar in general appearance (at least for positive x) to the quadratic proportion

$$y = Ax^2.$$

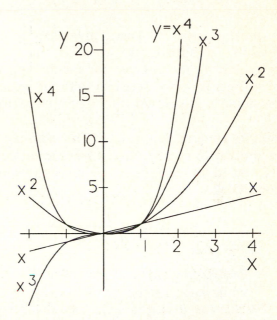

Each of these describes an increasingly steep monotonic rise in the value of y as x increases. Cubic (3rd power) and quartic (4th power) terms arise occasionally in introductory physics equations, but almost always as the only power term in the equation. In such cases the use of ratios in solving problems is a possibility, as in this example:

A spherical lead fishing weight of radius 0.50 cm has a mass of 5.9 g. How much lead is in a similar weight of radius 0.75 cm?

DISCUSSION. The amount of material in each weight is proportional to the volume which, for a sphere, goes as the cube of the radius R. (Specifically the formula is $V=(4/3)\pi R^3$.) Hence the ratio of the masses is the same as the ratio of the volumes which, in turn, is the same as the ratio of the radii cubed.

Using subscript 1 and 2 for the smaller and larger spheres respectively, we have

$$\text{Mass}_2 / \text{Mass}_1 = V_2 / V_1 = R_2^3 / R_1^3.$$

Thus

$$\text{Mass}_2 = \text{Mass}_1(R_2/R_1)^3 = (5.9 \text{ g})(1.5)^3 = 20 \text{ g}.$$

Solving problems using ratios should be considered whenever a physical quantity is known to be proportional to a power of some other quantity.

Thus for problems involving a simple "power law" $y = kx^n$, we can often find an answer even if the value of the coefficient k is not known.

INVERSE POWERS

In several places in physics you will come across functions whose form resembles these equations:

or

$$y = kx^{-1} = k/x$$

$$y = kx^{-2} = k/x^2 .$$

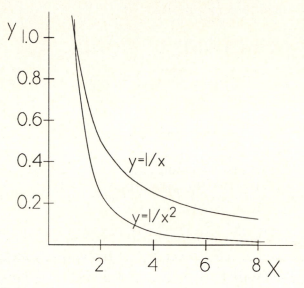

A major feature of such "inverse proportions" is that quantity y approaches zero as the quantity x gets extremely large. Likewise for smaller and smaller values of x the size of y increases without limit. This type of behavior is illustrated by the graphs on the right.

Simple inverse proportion. The example below, which is familiar to most students from their study of chemistry, has to do with the behavior of gases under ordinary conditions.

Boyle's law tells us that under most conditions if we change the volume V of a sample of gas while keeping the temperature constant, the pressure p of the gas will vary inversely as V. This can be expressed

$$p = k/V .$$

Suppose some ordinary room air (at atmospheric pressure p_a) is forced into a balloon where it occupies one third the volume it had outside the balloon. What is the pressure p_b inside the balloon compared to atmospheric pressure?

DISCUSSION. Set up the following ratio relationship:

or

$$\frac{P_b}{P_a} = \frac{k/v_b}{k/V_a} = \frac{V_a}{V_b} = \frac{V_a}{V_a/3} = 3$$

$$P_b = 3 P_a.$$

The pressure in the balloon is 3 times atmospheric pressure.

Inverse square laws. Phenomena which can be described by an equation of the type $y=kx^{-2}$ occur in several places in physics, almost always connected with some effect which decreases uniformly in all directions as the distance from a central point is increased. For example, the light intensity (a measure of the energy of the light illuminating an object) decreases as the inverse square of the distance from a light bulb. This phenomenon is used in the next example problem.

A lamp emits light equally in all directions. The intensity of the light illuminating an object at a distance r is given by $I = F/r^2$, where F is a constant related to the total energy of the light given off by the bulb. If the intensity reaching an object placed at 0.30 meters is 100 lux, what is the intensity at 1.0 meters?

DISCUSSION: To solve the problem it is not necessary to know the value of F, or even how the units of I are defined. The ratio expression is

$$\frac{I \text{ at } 1.0 \text{ meter}}{I \text{ at } 0.30 \text{ meter}} = \frac{(0.30 \text{ m})^2}{(1.0 \text{ m})^2}$$

which can be rearranged to give the answer:

$$\text{Intensity at } 1.0 \text{ meter} = (100 \text{ lux})(0.30/1.0)^2$$
$$= 9.0 \text{ lux}.$$

A "geometrical" interpretation of the inverse square. Electrical forces due to bits of electrical charge or gravitational forces due to concentrations of mass are two other situations in which some quantity varies as the inverse square of the distance from a central point. Such phenomena lend themselves to a common interpretation which is useful for visualizing the physics. In this "geometrical" interpretation we imagine that the central point "radiates" its influence in the form of a equally spaced straight lines emerging radially. Consider several concentric spheres located at distances r_1, r_2, etc. from the central point, as shown in this drawing. Since the surface areas of the spheres are larger the greater their radii, the number of lines that cross a given area on each sphere decreases with radius; i.e., the lines become less densely packed with distance. In fact since the surface areas increase as r^2, the "density" of the lines crossing each sphere shown in the drawing is proportional to $1/r_1^2$, $1/r_2^2$, etc. Thus we can picture "inverse-square phenomena" this way:

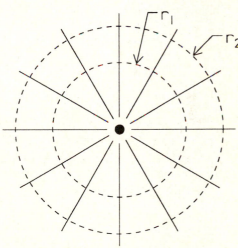

> • *a quantity which varies as the inverse square of the distance from a point is often represented by imaginary lines directed uniformly outward from the point; the density of these lines is proportional to the quantity at that distance.*

The lines are given different names when this idea is used in different parts of physics. For instance, in the example of light intensity given in the last example problem, the lines are often called "rays;" in electrical problems they are called "field lines."

FRACTIONAL POWERS

Powers and roots. Raising a number to a rational fraction (½, ⅓, etc.) is the reverse operation to taking an integer power. For example, squaring a square root gives back the original number, i.e., $(\sqrt{x})^2 = (x^{1/2})^2 = x$. (Note that the word "root" is used in a somewhat different sense than when it is used to denote the solutions of an equation.) Thus it is not surprising that the graphs of the square root and cube root shown on the right contrast strongly with graphs of integer power functions. Instead of more and more rapid increases of y as x gets larger, the increases in these functions "slow down" at large x.

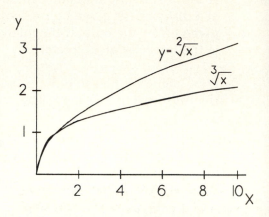

An example using ratios. The following is an example taken from physics which involves a fractional power function.

The time τ which a pendulum takes to make a complete swing back and forth is given by the following expression:

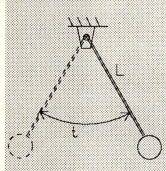

$$\tau = 2\pi\sqrt{\frac{L}{g}}$$

In this equation L is the length of the thin rod supporting the pendulum bob and g is the acceleration due to gravity. (Acceleration of gravity is discussed in Review 13.)

If a certain pendulum makes one swing in 1.0 s, how long will it take to make a swing if the supporting rod is shortened to half its original length?

DISCUSSION: The ratio approach eliminates the need to know most of the quantities in the equation. Thus we can write

$$\frac{\tau(\text{final})}{\tau(\text{original})} = \frac{\sqrt{L/2}}{\sqrt{L}}$$

$$\tau(\text{final}) = (1.0 \text{ s})\sqrt{\frac{1}{2}} = 0.71 \text{ seconds.}$$

Skill Drill 14

This drill covers the use of ratios or graphs in working problems involving various power functions. All the word problems are drawn from a science context.

1. Practice with ratios: For the following power functions, what is y when x = 8, given that y = 2 when x = 2?

(a) $y = kx^3$ (b) $x = ky^3$ (c) $y = kx^2$ (d) $y = A(kx)^{½}$

2. Boyle's Law (see example problem in Review 14) states that under most conditions the pressure p of a gas kept at constant temperature varies inversely with the volume V, i.e., $p = k/V$. Here is a table of data giving p and V for a sample of carbon dioxide gas at 295 K (approximately room temperature).

p(atmospheres)	V(liters)
0.5	112
0.8	68
1.0	54
1.5	37
1.8	30

(a) Plot these data on the p versus V graph at the right, and connect the points by a smooth curve. (b) Plot the data again on the p versus 1/V graph. (c) Draw a straight line which best fits the points plotted in the latter graph. From the slope of the line determine the constant k.

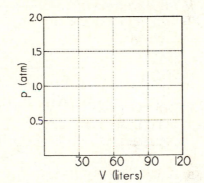

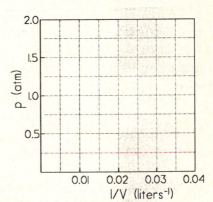

3. The attractive gravitational force exerted by the earth on an object located above the earth's surface depends as the inverse square on the distance of the object from the center of the earth. What is the pull of the earth on a satellite orbiting 26,000 mi above the surface, if the satellite weighs 1.0 ton when it is on the ground? The radius of the earth is 3960 mi.

4. Electrically charged bits of matter exert forces on one another which vary as the inverse square of their distance of separation r. Suppose, for a pair of electrical particles, the force F is written

149

Skill Drill 14 — SOLUTIONS AND ANSWERS

1. Practice with ratios: For the following power functions, what is y when x = 8, given that y = 2 when x = 2?

Let desired value = y'.

(a) $y = kx^3$

$$\frac{y'}{2} = \left(\frac{16}{2}\right)^3$$
$$y' = 2(8)^3$$
$$= 1024$$

(b) $x = ky^3$

$$16/2 = (y'/2)^3$$
$$y' = \sqrt[3]{8(8)}$$
$$= 4$$

(c) $y = kx^{-2}$

$$\frac{y'}{2} = \left(\frac{2}{16}\right)^2$$
$$y' = 2(4)/256$$
$$= 1/32 = 0.031$$

(d) $y = A(kx)^{1/2}$

$$\frac{y'}{2} = \sqrt{\frac{16}{2}}$$
$$y' = 4/\sqrt{2}$$
$$= 2.8$$

2. Boyle's Law (see example problem in Review 14) states that under most conditions the pressure p of a gas kept at constant temperature varies inversely with the volume V, i.e., p = k/V. Here is a table of data giving p and V for a sample of carbon dioxide gas at 295 K (approximately room temperature).

p(atmospheres)	V(liters)	$1/V$ (ℓ^{-1})
0.5	112	0.009
0.8	68	.015
1.0	54	.019
1.5	37	.027
1.8	30	.033

(a) Plot these data on the p versus V graph at the right, and connect the points by a smooth curve. (b) Plot the data again on the p versus 1/V graph. (c) Draw a straight line which best fits the points plotted in the latter graph. From the slope of the line determine the constant k.

$$k = slope = \frac{(1.7-0.6)\,atm}{(0.03-0.01)\,\ell^{-1}} = 55\,\ell\text{-}atm$$

3. The attractive gravitational force exerted by the earth on an object located above the earth's surface depends as the inverse square on the distance of the object from the center of the earth. What is the pull of the earth on a satellite orbiting 26,000 mi above the surface, if the satellite weighs 1.0 ton when it is on the ground? The radius of the earth is 3960 mi.

$$\frac{F'}{F} = \frac{R^2}{(R+h)^2} \quad \text{where } F \text{ is weight at surface.}$$

Thus
$$F' = (1\,T)\left(\frac{3960}{3960+26000}\right)^2 = 0.017\,T$$

4. Electrically charged bits of matter exert forces on one another which vary as the inverse square of their distance of separation r. Suppose, for a pair of electrical particles, the force F is written

$$F = \pm K/r^2$$

where the plus sign signifies a repulsive (outward directed) force which occurs when the charges on the two particles are of the same type ("positive" or "negative" charge), and the minus sign signifies an attractive (inward directed) force which occurs when the charges are of opposite type.

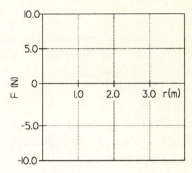

(a) On the graph at the right, plot F (in Newtons, N) vs. r for K = 9.0 N·m², for both attractive and repulsive forces. (b) At what separation is the force one hundred times greater than that exerted at 1.0 meter? (c) At what separation is the force one-hundredth as large as that exerted at 1.0 meter?

5. Stefan's law of energy radiation from hot bodies can be written

$$W = kT^4$$

where W is the intensity of the energy radiated per unit time and T is the absolute temperature (in Kelvin, K) of the surface.

(a) The sun's surface temperature is about 6000 K. If the sun cooled by just 1000 K, by what factor would the energy radiated to the earth be reduced?
(b) The earth is presently 93 million miles from the sun. How much closer would it have to be to receive its present solar radiation from a 5000 K sun? (Intensity of the radiation varies as the inverse square of the distance, as in the example problem in Review 14 on light intensity.)

6. The frequency υ (hertz *or* vibrations per second) of a stretched string varies as the square root of the force F with which it is stretched, i.e.,

$$\upsilon = k\sqrt{F}.$$

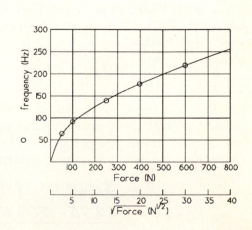

The graph shows some data for a steel piano string. Make another plot of the frequency data versus $\sqrt{F}$. Draw a straight line through these points to verify the square root dependence. From this line determine the constant k.

$$F = \pm K/r^2$$

where the plus sign signifies a repulsive (outward directed) force which occurs when the charges on the two particles are of the same type ("positive" or "negative" charge), and the minus sign signifies an attractive (inward directed) force which occurs when the charges are of opposite type.

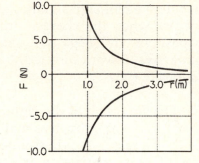

(a) On the graph at the right, plot F (in Newtons, N) vs. r for K = 9.0 N·m^2, for both attractive and repulsive forces. (b) At what separation is the force one hundred times greater than that exerted at 1.0 meter? (c) At what separation is the force one-hundredth as large as that exerted at 1.0 meter?

(a)

r (m)	F (N)
1	±9
1.5	±4
2	±2.3
3	±1

(b) $\dfrac{100\cancel{F}}{\cancel{F}} = \left(\dfrac{1.0\,m}{r}\right)^2 \longrightarrow r = \dfrac{1.0\,m}{\sqrt{100}} = 0.10\,m$

(c) $r = \dfrac{1.0\,m}{\sqrt{0.01}} = 100\,m$

5. Stefan's law of energy radiation from hot bodies can be written

$$W = kT^4$$

where W is the intensity of the energy radiated per unit time and T is the absolute temperature (in Kelvin, K) of the surface.

(a) The sun's surface temperature is about 6000 K. If the sun cooled by just 1000 K, by what factor would the energy radiated to the earth be reduced?

(b) The earth is presently 93 million miles from the sun. How much closer would it have to be to receive its present solar radiation from a 5000 K sun? (Intensity of the radiation varies as the inverse square of the distance, as in the example problem in Review 14 on light intensity.)

(a) $w'/w = \left(\dfrac{5000}{6000}\right)^4 = 0.48$

(b)

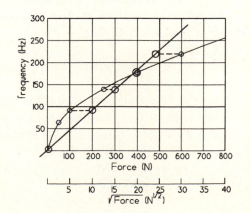

Move inward to distance r' where energy received is E'. At present distance r, energy E = 0.48 E'. Hence

$\left(\dfrac{r'}{r}\right)^2 = \dfrac{0.48 E'}{E'} = 0.48$

or

$r' = r\sqrt{0.48}$
$= (93 \times 10^6 \, mi) \sqrt{0.48}$
$= 64 \times 10^6 \, mi$

6. The frequency υ (hertz *or* vibrations per second) of stretched string varies as the square root of the force F with which it is stretched, i.e.,

$$\upsilon = k\sqrt{F}.$$

The graph shows some data for a steel piano string. Make another plot of the frequency data versus $\sqrt{F}$. Draw a straight line through these points to verify the square root dependence. From this line determine the constant k.

$k = $ slope of straight line

$= \dfrac{(275-50)\,Hz}{(30-5)\,N^{1/2}} = 9.0\,Hz\text{-}N^{-1/2}$

Fourth Round Posttest — Optimum test time: 30 minutes or less

This test is for you to assess the improvement in your skills using quadratic and other power functions. Some of the word problems have a physics-like context, as in some of the example problems and drill questions.

Have scratch paper and a calculator ready before you begin. Then time yourself and check your answers against those given after the last question.

STARTING TIME _____

ANSWERS

1. Solve the following equation and check your result by substituting the solutions back into the original expression:

$$\frac{3(x+1)}{5x-1} = \frac{2}{x} .$$

_____

2. Solve the following pair of simultaneous equations:

$$y = (x - 3)^2 \quad and \quad y = 3x^2 + 7x + 2 .$$

$(x,y)=$_____

$(x,y)=$_____

3. The hypotenuse of a right triangle is 2 cm longer than the greater of the sides. If that side is 7 cm longer than the shorter side, what are the dimensions of the triangle?

$(a,b,c)=$_____

_____

4. An equation from optics which relates the so-called "reflectivity" r of a piece of glass to the fraction of light energy transmitted T is as follows:

$$T = \frac{4r}{(1 + r)^2} .$$

If 90% of the light energy is transmitted, what are possible values of reflectivity r?

_____

5. A certain car, which is going 15 m/s, begins to accelerate at a constant rate just as it passes a traffic light at time t = 0. After that, its position x, as measured from the city center, is given by this equation:

$$x = (3.0 \text{ m/s}^2)t^2 + v_o t + 200 \text{ m} .$$

(a) How far from the city center is the traffic light?

_____

(b) What is the value of v_o?

_____

(c) What is the value of the acceleration?

_____

(d) How long does it take (after t=0) for the car to be 500 m from the city center?

_____

153

(e) On the graph, sketch a curve representing the motion after t = 0. How would you describe this type of curve?

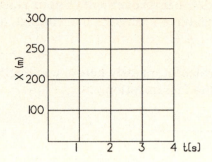

6. The electrical force of attraction between a "negative" and a "positive" charge varies as the inverse square of the separation distance between them. Written as an equation

$$F = -K/r^2 .$$

In the Bohr model of the hydrogen atom, an electron (negative) and a proton (positive) are separated by 0.0529 nm when the atom is in its lowest energy state. The force of attraction between the two particles, given by the equation above, is 8.2×10^{-8} N. What is the attractive force when the atom is in its next highest state, for which the separation distance is 0.2116 nm?

ENDING TIME_____

ANSWERS:

1. 2, ⅓

6. 0.51×10^{-8} N

2. (x,y) = ½, 6¼

   (x,y) = 7, 16

3. 8 cm, 15 cm, 17 cm

4. 1.9, 0.52

5. (a) 200 m

   (b) 15 m/s

   (c) 3.0 m/s²

   (d) 6.0 s

   (e) 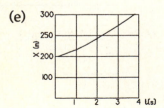   The curve is a parabola.

ESSAY: Energy and Other Conservative Issues

A great deal can be learned about how physics goes about its business from considering a simple helical spring. Fasten one end to a rigid support and attach a weight to the other end. Then, to eliminate the complications of gravity and rubbing, lay this system on its side with the weight supported on a "frictionless" table. [See drawing (a) below.] This "spring-mass oscillator" can be approximated very closely—although never exactly—in the laboratory. But it exemplifies a favorite ploy of physics: by thinking about such an "ideal" system we gain useful insight into how real things behave.

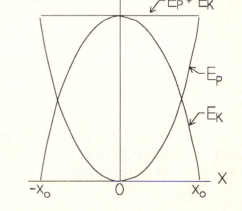

(a) (b) (c)

Now stretch the spring (spring constant k) and the attached weight (mass m) out a distance x_0. [See drawing (b).] The spring-mass system has been changed in a very significant way: now the mass has the potential, by virtue of the stretch in the spring, to acquire motion, if released. In fact, the released mass will move back to its original position; in doing so it will acquire a speed v_0. [See drawing (c).] Now the system is different in yet another way: the stretch of the spring is gone, but the mass will move by virtue of its inertia. The mass will overshoot its original position until the compressive force of the spring brings it to a stop. Then the mass will reverse its motion and head back towards the original central position. As time goes on *the mass will move back and forth, continually oscillating between extreme positions where the spring force dominates the action, and the central position where inertia is crucial.* This behavior has analogies in all kinds of oscillating phenomena in nature.

What can we say which is quantitative about the spring-mass oscillator? Clearly, the greater the initial stretch x_0, the greater the speed v_0 at the central position of the motion; in fact x_0 and v_0 are directly proportional. But to determine how the speed and the position of the mass are related at intermediate positions we call into play one of the more crucial discoveries of 18th century physics, viz., the importance for many dynamical systems of quadratic functions of speed v and position x. On the same graph at the right parabolas are plotted representing $\frac{1}{2}kx^2$ (the so-called *"potential energy"* E_p of the system) and corresponding values which one would measure of $\frac{1}{2}mv^2$ (called *"kinetic energy"* E_k). The remarkable thing is that when these two quantities are added to one another, the resulting value (*"total energy"* E_p+E_k) doesn't change at all during the oscillating motion. We express this by saying that the total energy is "conserved." *We can describe the spring-mass oscillator as an energy conserving system which continually and periodically transforms potential energy into kinetic energy and back again.*

The value of this discovery of how to describe "conservative" mechanical systems in terms of energy functions, is that once a value of the total energy is known, it is not necessary to bother with details of the dynamics (forces, Newton's laws, etc.) to find a relationship between the variables such as speed and position. But of course the world doesn't consist of ideal systems only.

Some have friction; in some the structural details are hidden. Nevertheless the idea of energy conservation is still of enormous value. We need only to recognize that there are other forms of energy, the sum of which remains constant within the boundaries of the system, i.e., is conserved, during all changes. For instance, for a mass sliding along a surface with friction we can take into account the associated heating effects—the trick is to know just what combination of measurements to call "heat energy."

But, at the heart of it, just what is energy? The late Richard Feynman, a great and frequently entertaining Nobel laureate in physics, constructed an allegory to help understand the answer to this question. He told of a mythical and mischievous boy named Dennis who delights in a game he plays with his scrupulous mother. Each day his mother counts Dennis' toy blocks, always finding the same number. To test his mother's vigilance, Dennis finds ways to hide the blocks: sealing them in a toy chest, sinking them in a basin of dirty water, etc. With matching cleverness, his mother devises ways to measure the number of unseen blocks: weighing the toy chest, checking the rise in the level of the dirty water, etc. Always the total of the number of seen blocks and unseen equivalent blocks remains unchanged. The analogies are clear: the number of blocks is the total amount of energy in a system, Dennis is Nature presiding over transformations from one form of energy to another, and Dennis' mother is science, which must find ways to measure and calculate the equivalent amounts of energy in its various forms. In the light of this allegory we recognize that fundamentally *energy can be regarded as a numerical quantity, measurable in its various forms in units equivalent to those of mv^2 (or kx^2), whose total for a complete physical system remains exactly the same despite any changes.* The essence of energy is that it is a conserved quantity.

Conservation is a powerful idea which is not restricted to energy. Physics relies on a number of conservation principles to guide its predictions. One of the most important conserved quantities is *momentum* (mv). For example, if two particles are hurled together in an atom smasher to form new particles, the total amount of momentum of each of the colliding particles added together must be the same as the total amount of momentum of all the newly created particles. Moreover, if the colliding particles are electrically neutral the total charge of all the resulting particles must add up to zero (conservation of charge principle). Some esoteric conservation laws may also apply. For instance, if among the colliding particles there are baryons (protons and some other related particles are called baryons), there must be just as many baryons among the resulting particles (conservation of baryon number).

This last mentioned conservation principle (of baryon number) can be used to illustrate how conservation laws act like a logical sieve which selects among all imaginable alternatives. For example, it "explains" why protons never break up spontaneously into several smaller particles: since a proton (1 baryon number) is the lightest baryon, there could be no baryons among the necessarily lighter daughter particles resulting from a breakup. The breakup is prohibited, otherwise baryon number would not be conserved. Nature plays a game with endless possibilities, but the moves of the game are bound by a strict set of rules which are very succinctly expressed in a relatively few laws of conservation.

Round V — Non-algebraic Functions

Previous Rounds considered physical phenomena whose behavior could be mathematically summarized using power functions, i.e., expressions in which relationships among variables can be written in terms of basic "arithmetic" operations: multiplication, division, addition, and subtraction. However the world abounds in events which are best described using "transcendental functions," relationships which traditionally have required numerical tables to determine. Today, these non-algebraic functions are readily evaluated at the touch of a button using complicated algorithms built into calculators and computers. The most important of these mathematical expressions are sinusoidal, exponential, and logarithmic functions.

PRETEST — Optimum time 15 minutes or less.

Except where the question indicates otherwise, use the special function keys on your calculator to evaluate trig functions, logarithms and e^x. Space is provided for your answers. As in previous pretests, keep track of your time, and finish the whole test before checking the answers which follow.

STARTING TIME_____   ANSWERS

1. On the graph at the right, a curve of the function
$y = A \sin kt$ has been drawn.

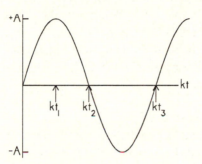

   (a) What are the values (in degrees) of kt_1, kt_2, and kt_3?

_____

   (b) What are kt_1, kt_2, and kt_3 in radians?

_____

2. (a) On the same graph used above to plot $y = A \sin kt$, sketch in a curve of $y = A \cos kt$.

   (b) If $k = \pi/4$ radians/s and $A = 2$ cm, evaluate $y = A \cos kt$ at $t = 3$ seconds.

_____

3. Water is being pumped into a closed air-filled tank (a "surge tank"). The more water which enters the tank, the higher the pressure P inside the tank, and the lower the rate of filling R. The result is an exponentially declining rate of filling the tank

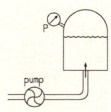

$$R = R_o\, e^{-t/T},$$

where $R_o = 10$ gal/min and T is a characteristic time for the filling process.
   (a) What is R at $t = T$?

_____

   (b) If $T = 10$ min, how long does it take from $t = 0$ until $R = 1.4$ gal/min?

_____

157

4. Without a calculator find numerical values for the following:

$\log_2 1$ _____

$\log_{10} 0.1$ _____

$\ln e^2$ (ln is the logarithm to the natural base e.) _____

5. The acidity of a water solution is stated in terms of a "pH" scale defined by

$$pH = \log_{10}\frac{1}{[H^+]}$$

where [H+] is the number of moles of hydrogen (more correctly, hydronium) ions per liter of solution. Written in scientific notation, what are the hydronium ion concentrations (mol/l) in the following solutions?

(a) household ammonia (pH = 12); _____

(b) lemon juice (pH = 2). _____

ENDING TIME_____

ANSWERS:

1. (a) 90°, 180°, 360°
   (b) $\pi/2$, π, 2π rad

2. (a)
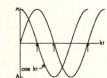
   (b) -2 cm

3. (a) $0.37R_o = 3.7$ gal/min
   (b) $t = 2T = 20$ min

4. 0, -1, 2

5. (a) 10^{-12} mol/l
   (b) 10^{-2} mol/l

Review 15 — Sinusoidal Functions: Oscillating Phenomena

Much of what we casually observe around us changes in a cyclic or repeating way: the coming and going of seasons, the rise and fall of water along the edge of a beach, the beating of a butterfly's wings. Periodic oscillations are of particular importance in the realm of very rapid and microscopic changes and are intimately connected with moving waves: sound consists of vibrations in the air and light depends on the transmission of periodically varying electric and magnetic forces. Fundamental to the quantitative treatment of such phenomena is the use of sinusoidal functions, expressions in which the independent variable is contained in the argument of a sine or cosine. This Review goes over basic ideas needed to use these functions in connection with oscillations and waves, and also discusses radian angle measure and its applications.

SINUSOIDAL FUNCTIONS

Describing oscillations mathematically. Many important physical systems —such as bouncing springs or swinging pendula — are observed to oscillate in a way which can be simply described using the sine and cosine functions first encountered in your study of trigonometry. Even systems which exhibit more complicated repetitive motion can be analyzed using expressions which are combinations of simple sine and cosine functions.

In trigonometry the quantity θ (the "argument") in $\sin \theta$ and $\cos \theta$ is usually regarded as a dimensionless *constant*, representing some fixed direction in terms of degrees. When used for describing oscillating behavior the argument instead *varies* uniformly with time t: $\theta = kt$. As t increases the argument goes progressively from 0° through all intermediate values to 360° (which is equivalent to 0°) and continues to repeat this sequence; the corresponding values of sine and cosine oscillate continuously between +1 and -1. Thus the expressions

$$x = A \cos kt \quad and \quad y = A \sin kt$$

both represent quantities which vary continuously and periodically between A and -A. The units of x and y are the same as that of A; if t is measured in seconds, k is in deg/sec. It is also common practice to express the argument in radians, as explained later in this Review, but for the time being we will use degrees.

The following problem uses the "spring-mass oscillator" discussed in the previous Essay, as an example of an oscillating system whose motion can be described using a sinusoidal function.

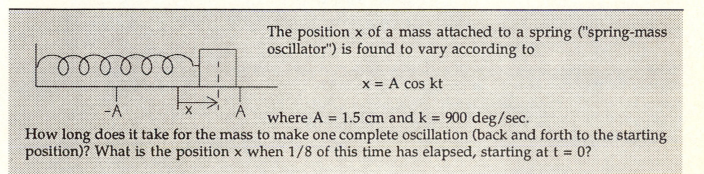

The position x of a mass attached to a spring ("spring-mass oscillator") is found to vary according to

$$x = A \cos kt$$

where A = 1.5 cm and k = 900 deg/sec.

How long does it take for the mass to make one complete oscillation (back and forth to the starting position)? What is the position x when 1/8 of this time has elapsed, starting at t = 0?

DISCUSSION: A complete oscillation requires the argument of the cosine to change by 360°. Thus kt = 360°, or

$$t = 360°/k = 0.40 \text{ seconds.}$$

One eighth of a full oscillation takes (0.40 sec)/8 = 0.050 sec. The argument of the cosine is kt = (900 deg/sec)(0.050 sec) = 45°. Thus
$$x = (1.5 \text{ cm}) \cos 45° = 1.1 \text{ cm.}$$

Curves of sine and cosine. Sin θ and cos θ versus θ are plotted one under the other at the right. Both have the same "sinuous" shape, but are shifted relative to one another along the θ-axis by 90°. You should be able readily to sketch these curves.

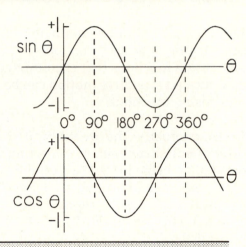

Sin θ and cos θ vary continuously between the extreme values ±1 at intervals of θ = 360°, crossing through the axis with a finite slope; sin θ rises at θ = 0 from zero towards a maximum, whereas cos θ has a maximum at θ = 0.

These guidelines are applied in the following example exercise:

On a graph of position x versus time t sketch a curve showing the motion (for one complete oscillation) of the "spring-mass oscillator" described in the last example.

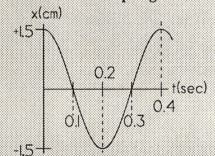

DISCUSSION: The motion is described by a cosine function:

$$x = (1.5 \text{ cm}) \cos (900°/\text{sec})t .$$

At t=0 the curve has a maximum x = 1.5 cm. It returns to that value when the argument (900 °/sec)t = 360° , i.e., when t = 0.40 sec, as shown.

Phase angle. Whether a sinusoidal oscillation is to be described by cos kt or sin kt clearly depends on whether t = 0 is chosen to be at the maximum or at the zero-point of the oscillation. A more general choice for t = 0 can be expressed by including a constant θ_o, called a *phase angle*, in the argument of the function. Thus, the curves of the functions

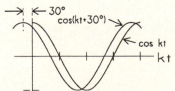

$$x = A \cos(kt+\theta_o) \quad and \quad y = A \sin(kt+\theta_o)$$

resemble the cos kt and sin kt curves, but are shifted along the kt-axis by an amount θ_o. A cosine curve with $\theta_o = +30°$ is pictured here.

A positive phase angle θ_o shifts the sinusoidal curve by an amount θ_o towards lower values of kt; a negative phase angle shifts the curve towards higher values.

Trigonometric identities. Not all repeating phenomena can be described by simple sine or cosine functions. However, at least in principle, all periodic functions can be formed from combinations of simple sines or cosines. Some especially important cases require the combining of just two sinusoidal functions into a more valuable expression, often in a form in which the several arguments are joined together.

A number of formulas (trig identities) for combining two or more sinusoidal functions have been worked out and are listed in trigonometry text books. If such mathematics is needed in a physics problem, the identity will usually be given or you will be able to look it up. It is mostly important that you be aware that the identities exist. Here are a couple of examples:

$$\sin \theta + \sin \phi = 2 \sin \tfrac{1}{2}(\theta + \phi) \cdot \cos \tfrac{1}{2}(\theta - \phi)$$

$$\sin \theta \cdot \cos \phi = \tfrac{1}{2} \sin (\theta + \phi) + \tfrac{1}{2} \sin (\theta - \phi)$$

RADIAN MEASURE

Arcs and angles. In the analysis of periodic phenomena using radians as the units of angle measure, rather than degrees, generally leads to simpler formulas. This is because the size of a degree is based on an arbitrary choice, viz., the choice of 360° to represent one complete rotation. Radian measure is a more natural system, based on the geometry of a circle, as follows:

The size of an angle, measured in radians, is the ratio of the length of the circular arc subtended by that angle to the distance to the arc.

In terms of the diagram at the right

$$\theta = S/R \text{ , } \theta \text{ in radians.}$$

Whereas radian measure is often preferred in the treatment of oscillating systems, degrees are more often used to describe direction (especially in experimental situations). However, there is some value in being able to use either system of measurement in both types of work. For instance, in a situation in which an arc length is of importance, radians may be favored since the simple formula $S = R\theta$ can be used.

Interchanging radians and degrees. Many scientific calculators have DEG/RAD conversion functions. But apart from that, there is little need to memorize a conversion factor; one need only remember that a full circle is an arc of length $2\pi R$, and corresponds to $(2\pi R)/R = 2\pi$ radians. Thus the direct proportion relating the units can be stated

• *angle in radians is to angle in degrees as 2π is to 360.*

Certain cardinal point conversions are important enough to remember outright. As illustrated in the following problem, these include

$$45° = \pi/4 \text{ radians} \qquad 90° = \pi/2 \text{ radians}$$
$$180° = \pi \text{ radians} \qquad 270° = 3\pi/2 \text{ radians} .$$

These angles are used in labeling the drawing in the following problem.

Write an equation for the motion of the spring-mass oscillator described in the previous examples using radians. Sketch a graph of the motion and label several significant points on the time axis with the corresponding values of radians.

DISCUSSION: The coefficient k in the argument is given in terms of radian measure as

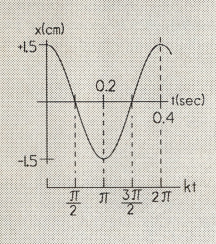

$$k = \left(900 \; \frac{\text{deg}}{\text{rad}} \right)\left(\frac{2\pi \; \text{rad}}{360 \; \text{deg}} \right) = 5\pi \; \frac{\text{rad}}{\text{sec}}$$

The motion is thus given by

$$x = (1.5 \text{ cm}) \cos 5\pi t$$

where, according to common practice, units are left out of the argument. (Usually, the appearance of π explicitly in the argument signals that radians is the unit.) In the drawing, the points chosen for labelling are maxima, minima, and axis crossing points (corresponding to the spring at its unextended length.)

Small angle approximations. Angle θ in the right triangle pictured here is small compared with a right angle. The arc S subtended by θ at distance R is drawn in for comparison with the side of the triangle d opposite to θ. The two lines are nearly indistinguishable: $S \simeq d$. As an approximation, we can use S instead of d in the definitions of the trig functions of θ. For instance, $\sin \theta = d/R \simeq S/R = \theta R / R$ 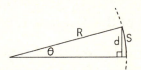 $= \theta$, which is the angle itself (providing θ is expressed in radians.) Expressions for other trig functions can likewise be simplified.

For small angles θ, when θ is expressed in radians: $\sin \theta \simeq \theta$, $\tan \theta \simeq \theta$, and $\cos \theta \simeq 1$.

These approximations are extremely good for angles of the order of degrees. For example, for $\theta = 5° = 0.09$ rad, $\sin \theta$ differs from the radian value by only about 0.1% .

A building viewed at a distance of 0.50 mile subtends an angle of 11.0°. Use a small angle approximation to estimate the building's height. Compare this result with the more exact value obtained using a trig function.

DISCUSSION: In radians, $\theta = (11.0°)(2\pi \text{ rad}/360°) = 0.192$ rad.

Using the small angle approximation

$$h/x = \tan \theta \simeq \theta = 0.192$$

or

$$h \simeq (0.192)(0.50 \text{ mi})(5280 \text{ ft/mi})$$
$$= 507 \text{ ft.}$$

More exactly, $h = x \tan 11° = 513$ ft, about 1.2% greater than the approximate answer.

Skill Drill 15

Two exercises in this drill should give you some insight into certain aspects of sinusoidal functions not explored in the Review: their connection with rotational motion and with wave phenomena. Other questions should help your understanding of the interrelationships among sinusoidal functions and the use of radian measure.

1. This is an exercise to help you visualize the shape of a sinusoidal curve as well as to demonstrate its relationship to rotational motion.

Consider the rotating wheel of radius R pictured below. A spot is painted on the rim at the end of a radial line which at any moment makes an angle θ with respect to the horizontal. The rotation is uniform and time t = 0 when the radial line is horizontal, so that θ = kt.

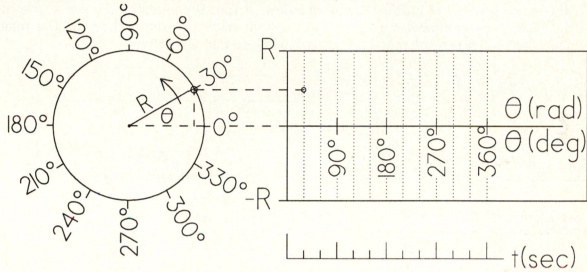

(a) Draw the radial line and spot at each of the marked angular positions from 0° to 360°. (The 30° position is shown in the drawing.) For each position plot the height of the spot above the axis versus angle by drawing with a ruler a horizontal line from the spot to the appropriate point on the graph.
(b) Connect the plotted points by a smooth curve; you will recognize this as the curve of sin kt. (c) Label the horizontal axis with equivalent radian values at each of the positions labeled above in degrees.
(d) Taking k = 2π, mark each of these positions on the t(sec) axis with the equivalent values of time.

2. A spring-mass oscillator moves back and forth between positions x = ±5.0 cm in a manner described by a sinusoidal function. A complete oscillation requires 1.0 second.

(a) Taking t = 0 when x = 5.0 cm, sketch a curve representing the motion on the accompanying graph.

(b) On the same graph sketch the curve of the motion if x = 0 when t = 0.

(c) There is an alternative curve which also satisfies the conditions stated in (b). Draw this curve as well.

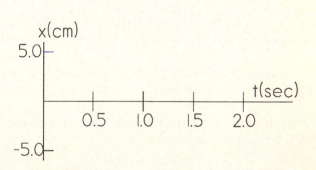

Skill Drill 15 — SOLUTIONS AND ANSWERS

1. This is an exercise to help you visualize the shape of a sinusoidal curve as well as to demonstrate its relationship to rotational motion.

Consider the rotating wheel of radius R pictured below. A spot is painted on the rim at the end of a radial line which at any moment makes an angle θ with respect to the horizontal. The rotation is uniform and time t = 0 when the radial line is horizontal, so that θ = kt.

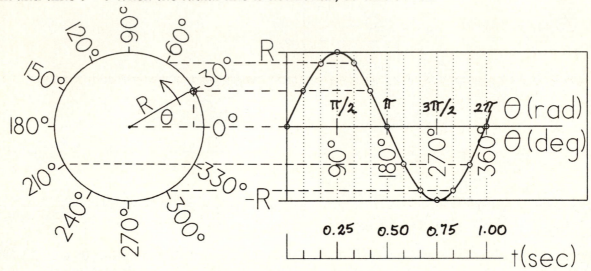

(a) Draw the radial line and spot at each of the marked angular positions from 0° to 360°. (The 30° position is shown in the drawing.) For each position plot the height of the spot above the axis versus angle by drawing with a ruler a horizontal line from the spot to the appropriate point on the graph. (b) Connect the plotted points by a smooth curve; you will recognize this as the curve of sin kt. (c) Label the horizontal axis with equivalent radian values at each of the positions labeled above in degrees. (d) Taking k = 2π, mark each of these positions on the t(sec) axis with the equivalent values of time.

$$2\pi t = \pi/2 \longrightarrow t = 0.25 \text{ s. etc. for each additional quarter rotation.}$$

2. A spring-mass oscillator moves back and forth between positions x = ±5.0 cm in a manner described by a sinusoidal function. A complete oscillation requires 1.0 second.

(a) Taking t = 0 when x = 5.0 cm, sketch a curve representing the motion on the accompanying graph.

(b) On the same graph sketch the curve of the motion if x = 0 when t = 0.

(c) There is an alternative curve which also satisfies the conditions stated in (b). Draw this curve as well.

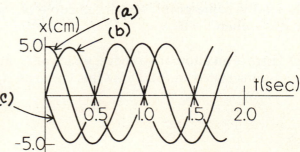

3. Beneath the end of a pier the level of the water is observed to move up and down as waves move by according to the equation

$$y = (10.\text{cm}) \cos (\pi/2)t ,$$

where y is the height of the water surface with respect to the average height. t is the time in seconds and the argument is in radians.

(a) How long does one oscillation take?

(b) What is the total distance the surface moves (highest to lowest point) during an oscillation?

(c) What is y at t = 2 sec?

4. On each of the graphs sketch curves of the indicated functions (arguments in radians):

(a) $x = x_o \sin (\theta + \pi)$.

(b) $x = x_o \sin (\theta - \pi/4)$.

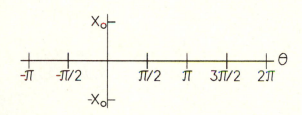

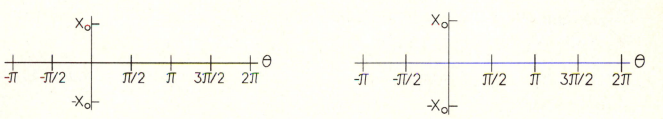

(c) $x = x_o \sin (\theta + \pi/4)$.

(d) $x = x_o \cos (\theta - \pi/2)$.

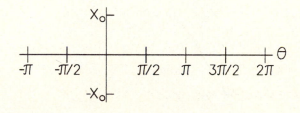

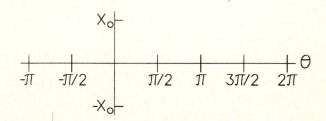

5. From an inspection of the curves drawn in the preceding question, determine the phase angle θ_o in each of the following identities:

(a) $\sin(\theta + \theta_o) = - \sin \theta$

(b) $\cos(\theta + \theta_o) = \sin \theta$

6. Verify each of the equations in the preceding problem by substituting into one of the following general trig identities:

$$\sin(\theta \pm \phi) = \sin \theta \cdot \cos \phi \pm \cos \theta \cdot \sin \phi$$
$$\cos(\theta \pm \phi) = \cos \theta \cdot \cos \phi \mp \sin \theta \cdot \sin \phi$$

3. Beneath the end of a pier the level of the water is observed to move up and down as waves move by according to the equation

$$y = (10.cm) \cos (\pi/2)t ,$$

where y is the height of the water surface with respect to the average height. t is the time in seconds and the argument is in radians.

(a) How long does one oscillation take?

(b) What is the total distance the surface moves (highest to lowest point) during an oscillation?

(c) What is y at t = 2 sec?

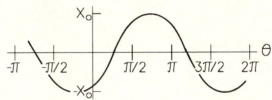

(a) For one oscillation
$$(\pi/2)t = 2\pi$$
or $t = 4s$

(b) Lowest: $y = -10cm$
   Highest: $y = +10 cm$
   Total: $20.0 cm$

(c) $y = (10.0cm) \cos (\pi/2)(2)$
     $= 10.0 cm \cos \pi$
     $= -10.0 cm$

4. On each of the graphs sketch curves of the indicated functions (arguments in radians):

(a) $x = x_o \sin (\theta + \pi)$.

(b) $x = x_o \sin (\theta - \pi/4)$.

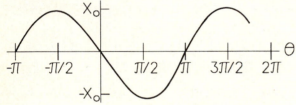

(c) $x = x_o \sin (\theta + \pi/4)$.

(d) $x = x_o \cos (\theta - \pi/2)$.

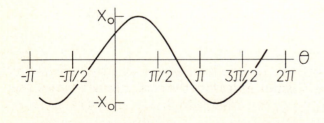

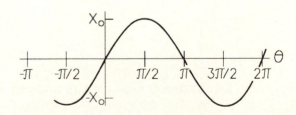

5. From an inspection of the curves drawn in the preceding question, determine the phase angle θ_o in each of the following identities:

(a) $\sin(\theta + \theta_o) = - \sin \theta$

From (a) above:
   $\theta_o = \pi$

(b) $\cos(\theta + \theta_o) = \sin \theta$

From (d) above:
   $\theta_o = -\pi/2$

6. Verify each of the equations in the preceding problem by substituting into one of the following general trig identities:

$$\sin(\theta \pm \phi) = \sin \theta \cdot \cos \phi \pm \cos \theta \cdot \sin \phi$$
$$\cos(\theta \pm \phi) = \cos \theta \cdot \cos \phi \mp \sin \theta \cdot \sin \phi$$

7. Travelling waves use mathematics similar to that used to treat oscillations: the wave profile along a direction x (such as the undulating surface of water) can be sketched out using a sinusoidal function with argument (k′x); a phase angle kt determines how the undulation is shifted along x with time. Consider, for instance, the height y of the surface in a water wave given by

$$y = (0.50 \text{ m}) \cos (k′x - kt)$$

where k′ = π/12 rad/m and k = π/2 rad/sec. Sketch curves giving a profile of the water surface at t = 0, 1 sec, and 2 sec. (HINT: After sketching the first curve, evaluate kt and shift the curve accordingly.)

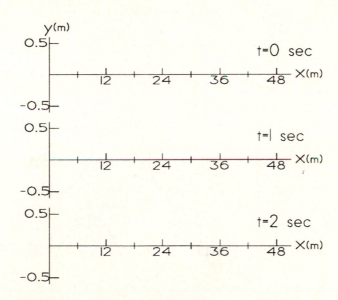

8. The diameters D of the sun and moon and their distances R from the earth are as follows: sun, D = 1.4 × 10⁶ km, R = 150 × 10⁶ km; moon, D = 3.5 × 10³ km, R = 380 × 10³ km. What angle does each of these astronomical bodies subtend when viewed from the earth — in radians? — in degrees?

9. An aircraft search light beam spreads out at an angle of 2.0°, illuminating a patch of clouds directly above at 10,000 ft elevation. What is the angular spread of the beam in radians? (Do not use a DEG/RAD conversion key on your calculator.) From this, estimate the width in feet of the beam at cloud level.

(a) Let $\phi = \pi$

$\sin(\theta + \pi) = \sin\theta \cos\pi + \cos\theta \sin\pi$
$= \sin\theta(-1) + \cos\theta(0)$
$= -\sin\theta$

(b) Let $\phi = \pi/2$

$\cos(\theta - \pi/2) = \cos\theta \cos\pi/2 + \sin\theta \sin\pi/2$
$= \cos\theta(0) + \sin\theta(1)$
$= \sin\theta$

7. Travelling waves use mathematics similar to that used to treat oscillations: the wave profile along a direction x (such as the undulating surface of water) can be sketched out using a sinusoidal function with argument (k'x); a phase angle kt determines how the undulation is shifted along x with time. Consider, for instance, the height y of the surface in a water wave given by

$$y = (0.50 \text{ m}) \cos(k'x - kt)$$

where k' = π/12 rad/m and k = π/2 rad/sec. Sketch curves giving a profile of the water surface at t = 0, 1 sec, and 2 sec. (HINT: After sketching the first curve, evaluate kt and shift the curve accordingly.)

Plot
$y = (0.5m) \cos \frac{\pi x}{12}$

Plot
$y = (0.5m) \cos\left(\frac{\pi x}{12} - \pi/2\right)$

Plot
$y = (0.5m) \cos\left(\frac{\pi x}{12} - \pi\right)$

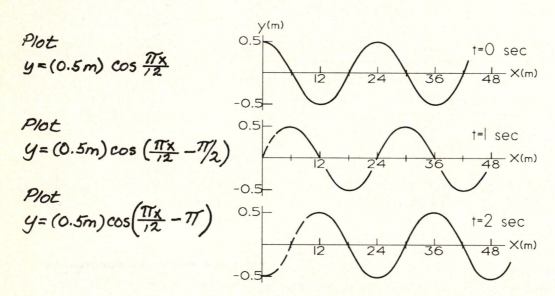

8. The diameters D of the sun and moon and their distances R from the earth are as follows: sun, D = 1.4 × 10⁶ km, R = 150 × 10⁶ km; moon, D = 3.5 × 10³ km, R = 380 × 10³ km. What angle does each of these astronomical bodies subtend when viewed from the earth — in radians? — in degrees?

Sun: $\theta = \frac{D}{R} = \frac{1.4\times10^6 \text{ km}}{150\times10^6 \text{ km}} = 9.2\times10^{-3} \text{ rad}$

$= 9.2\times10^{-3}\left(\frac{360°}{2\pi}\right) = 0.53°$

Moon: $\theta = \frac{D}{R} = \frac{3.5\times10^3}{880\times10^3} = 9.2\times10^{-3} \text{ rad}$

The moon almost exactly covers the sun during an eclipse!

9. An aircraft search light beam spreads out at an angle of 2.0°, illuminating a patch of clouds directly above at 10,000 ft elevation. What is the angular spread of the beam in radians? (Do not use a DEG/RAD conversion key on your calculator.) From this, estimate the width in feet of the beam at cloud level.

$\theta = 2.0°(2\pi \text{ rad}/360°) = 0.035 \text{ rad}$

Width of beam $= \theta h = (0.035)(10\,000 \text{ ft})$
$= 350 \text{ ft}$

Review 16 — Exponential Functions: Growth and Decay

The world abounds in situations in which some quantity rapidly grows in size — not only without apparent limit but also with ever increasing rapidity. We hear of runaway nuclear reactions or explosive population growth — phenomena which almost imply a lack of control. These are examples in which growth, in fact, feeds upon itself — the larger the quantity, the larger the rate at which it increases. Most of these situations, especially in physics, can be described using an exponential function like a^{bt} (t being time). Another group of important occurrences are described by the related exponential decay function a^{-bt}; in such situations quantities fade away towards zero at an ever declining rate. This Review deals with the essential mathematics needed to apply these functions to phenomena discussed in physics courses.

EXPONENTIAL GROWTH

The exponential function. The function of x

$$y = a^{bx}$$

is an exponential function with "base a." The exponent bx must be dimensionless.

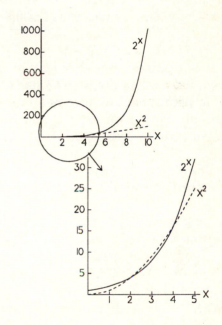

It is instructive to compare a function of this type, for instance 2^x (base 2), with a corresponding power function (x^2) for positive values of x. The two functions are plotted at the right; an expanded portion of the curves (at small x) is also shown. Despite a rapid rise of x^2 at smaller values of x, the function 2^x ultimately catches up and surpasses it. This characteristic behavior holds no matter what the base or power; the exponential function ultimately will overwhelm any power function. This is important because many physical phenomena meet the criterion for exponential growth, as follows:

A quantity for which the rate of growth is proportional to the quantity itself follows an exponential growth curve.

A non-physical example, compound interest, illustrates this idea:

A certain money account is credited with 10% interest every January 1. If this account starts out with a principal P_o, develop an expression for the principal P_t after t years. If P_o = \$100, what is the account worth, assuming no withdrawals, after 10 years? after 100 years?

DISCUSSION: After 1 year the account is worth $P_1 = P_o + 0.1P_o = P_o(1.1)$. After two years $P_2 = P_o(1.1) + (0.1)P_o(1.1) = P_o(1.1)^2$. Extending the reasoning to t years we get

$$P_t = P_o(1.1)^t .$$

This is an exponential growth equation with base (1.1). (The exponent, which is the number of times the base is to be multiplied by itself, must be dimensionless. Thus t should be interpreted as "number of interest periods" = time in years ÷ time in an interest period.)

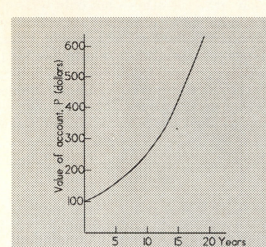

After 10 years

$$P_{10} = (\$100)(1.1)^{10} = \$259 .$$

After 100 years

$$P_{100} = (\$100)(1.1)^{100} = \$1,378,061.$$

Thus, even for a relatively weak exponential function (base close to 1), the potential for growth is inexorable, given sufficient time. The account values for the first 20 interest periods, joined by a smooth curve, are plotted at the left.

In this example the exponential is interpreted in its most basic way, as an integer number of multiplications of the base by itself. It requires more advanced mathematics to understand the meaning of the base raised to a general non-integer power. For practical purposes, however, it is sufficient to think of the general exponential function as the numbers along the smooth, continuous curve joining integer powers of base a. These values can be found using electronic calculators or computers.

Choice of a base. Certain bases lend themselves well to certain applications, even though an exponential expressed using one base can be rewritten in terms of any other base (see below).

For example, 2^x is handy for describing quantities which double in given intervals of time and for the discussion of probabilities. Moreover, integer values of x generate the numbers 2, 4, 8,... of the binary counting system used in computers. Base 10 is also convenient; for integer values of x, 10^x yields the place values 10, 100, 1000,... of our usual decimal counting system. There is, however, still another base lying between 2 and 10, which is often favored because it generally gives rise to simpler expressions in mathematical derivations involving exponential functions. This is the "natural base" e = 2.7182..., discussed below.

Natural base e. For any exponentially varying quantity $y = a^{bx}$, the rate at which y increases is proportional to y itself; the proportionality constant depends on the choice of base. The rate of increase can be estimated from a curve of y versus x; at any point on the curve the rate is approximately rise ÷ run = $\Delta y/\Delta x$. For example, the curve of 10^x and the corresponding curve of $\Delta y/\Delta x$ versus x are plotted at the right; a similar pair of curves for 2^x is also shown. In one case the rate exceeds the function itself; in the other case the rate is smaller than the function.

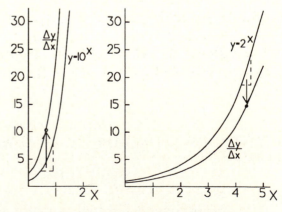

On the other hand, for the natural base e, the curves of $y = e^x$ and $\Delta y/\Delta x$ versus x are identical; the proportionality constant linking the two quantities is 1. This fact accounts for the mathematical simplicity inherent in using e as the base of the exponential function.

For the natural base e = 2.7182..., the rate of increase of e^x equals e^x itself.

The following example explores some of the salient features of the mathematical description of exponential growth which uses the natural base.

Suppose a population of animals p increases exponentially according to

$$p = p_o \, e^{t/T},$$

where t is in years, p_o = 1000 animals, and T = 0.50 year. Plot p as a function of t out to 1 year. What is the significance of the constant p_o? of the constant T? From the curve estimate the time it takes for the population to double and compare this with T.

DISCUSSION: The e^x key on your calculator can be used to plot the curve shown here. An example calculation is:

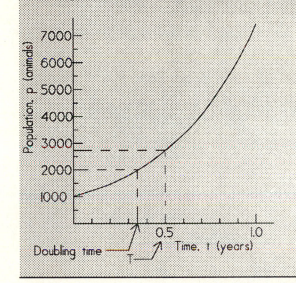

At t = 0.30 years

$$p = (1000 \text{ animals})e^{0.60}$$
$$= 1822 \text{ animals}.$$

The curve intersects the vertical axis at p_o = 1000 animals, which is the initial population (at time t = 0).

T is the length of time for the population to increase by a factor e, i.e., to 2718 animals. This *characteristic time* is somewhat greater than the *doubling time*, which, from the graph, is estimated to be about 0.35 years.

Transformation of bases. The example problem given above asks about "doubling time," which is sometimes a desirable way to think about certain exponential growth phenomena; during any period equal to a doubling time the quantity increases by a factor of 2. In such cases it may be preferable to describe growth using base 2, this way:

$$y = y_o \, 2^{t/T_2}$$

where T_2 is the doubling time. T_2 is proportional to T, the "characteristic time" used in the exponential function with base e. (See the example above.) In fact

• *any base can be used to describe exponential growth; the exponents used with the various bases are proportional to one another.*

The following formulas relating growth functions using different bases can be derived using logarithms. (See Review 17.)

$$2^x = e^{0.69\,x} \qquad\qquad 2^x = 10^{0.30\,x}$$

$$10^x = e^{2.3\,x} \qquad\qquad e^x = 10^{0.43\,x}$$

EXPONENTIAL DECAY

Just as there are cases of a quantity whose rate of growth is proportional to the quantity itself, a large number of phenomena exhibit *rates of decrease* of a quantity which are proportional to the quantity. These cases of *"exponential decay"* are most often written in terms of the natural base e; the exponent usually contains an explicit minus sign.

> *A quantity for which the rate of decrease is equal to the quantity itself is described by an exponential decay function $y = y_o e^{-x}$.*

As in the case of exponential growth, the exponent is dimensionless, and is often written as a ratio t/T, where T is often called the "time constant."

> *During one time constant an exponentially decaying quantity is reduced to $1/e$ (37%) of its former value.*

A graph typical of exponential decay is examined in the Discussion part of the following example problem.

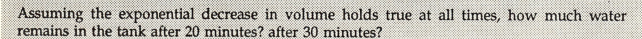

Water is flowing from a small hole in the bottom of a cylindrical tank. The rate of flow at any time depends on the volume V remaining in the tank, such that

$$V = V_o e^{-t/T} ,$$

where $V_o = 10$ gallons and $T = 10$ minutes.

What is the significance of V_o? of T? Plot V versus t out to 20 minutes and mark the positions of $V = V_o$ and $t = T$.

Assuming the exponential decrease in volume holds true at all times, how much water remains in the tank after 20 minutes? after 30 minutes?

DISCUSSION: V_o is the initial volume of water in the tank (at $t = 0$). T, the time constant of the decay, is the period of time it takes for the volume to decline to 0.37 of its value at the beginning of that period.

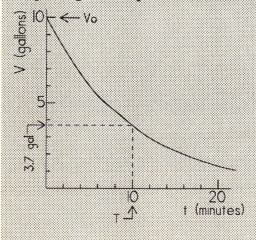

Thus, at $t = T = 10$ minutes

$$V = (0.37)(10 \text{ gal}) = 3.7 \text{ gal.}$$

Likewise, after another 10 minutes, i.e., at $t = 2T$

$$V = (0.37)(3.7 \text{ gal}) = 1.4 \text{ gal.}$$

Finally, after another time interval T (at $t = 30$ minutes)

$$V = (0.37)(1.4 \text{ gal}) = 0.5 \text{ gal.}$$

Skill Drill 16

Most of the exercises in this drill aim to reinforce your grasp of the fundamental character of the exponential function. Additional problems review the use of such functions to describe growth and decay in natural phenomena.

　　1. Construct a curve of y/y_o versus t/T using the following graphical approach:

(a) Starting on the vertical axis at $y/y_o = 1$, draw a line segment with slope = 1 out to $t/T = 0.25$. (This first step is done for you at the right.)

(b) Continue the curve by next drawing a line segment between $t/T = 0.25$ and 0.5 with a slope equal to y/y_o at $t/T = 0.25$.

(c) Add to the curve by successively drawing line segments for 4 more intervals of $t/T = 0.25$, each time with a slope equal to y/y_o at the start of the interval.

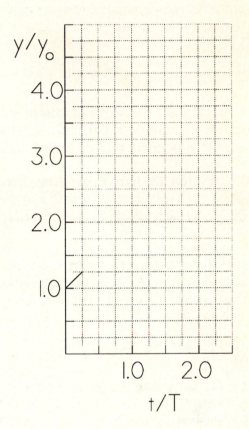

(d) On the same graph draw in several circled points calculated from the function $y/y_o = e^{t/T}$, and compare these with the approximation to an exponential curve just constructed.

　　2. (a) Plot a curve of $y = 10^x$ by evaluating y at $x = 0$, 1/4, 1/2, 3/4, and 1, and connecting the points with a smooth curve. (If your calculator does not have a 10^x key, use the $\sqrt{\ }$ key to calculate the fourth and other roots.)

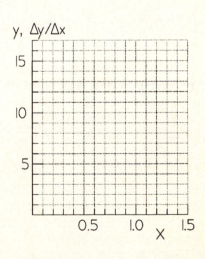

(b) Estimate the rates of increase of y at several points ($x = 0.2$, 0.4, 0.9) by drawing lines tangent to the curve at those points and calculating the slopes of the lines. Plot these rates of increase on the same graph.

(c) Compare the data you have plotted. What is the proportionality constant relating 10^x and its rate of growth?

173

Skill Drill 16 — SOLUTIONS AND ANSWERS

1. Construct a curve of y/y_o versus t/T using the following graphical approach:

(a) Starting on the vertical axis at $y/y_o = 1$, draw a line segment with slope = 1 out to $t/T = 0.25$. (This first step is done for you at the right.) *Mark segment #1.*

(b) Continue the curve by next drawing a line segment between $t/T = 0.25$ and 0.5 with a slope equal to y/y_o at $t/T = 0.25$. *Segment #2:*

Rise/Run = slope$_2$ $= \frac{y}{y_o} = 1.25$

(c) Add to the curve by successively drawing line segments for 4 more intervals of $t/T = 0.25$, each time with a slope equal to y/y_o at the start of the interval.

slope$_3$ = 1.5
slope$_4$ = 1.8
slope$_5$ = 2.4
slope$_6$ = 3.1

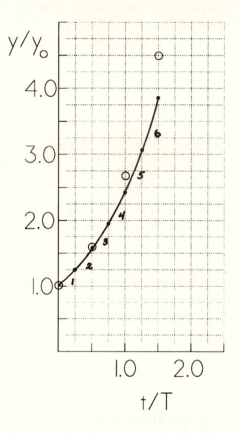

(d) On the same graph draw in several circled points calculated from the function $y/y_o = e^{t/T}$, and compare these with the approximation to an exponential curve just constructed. *For $t/T = 0, 0.5, 1.0, 1.5$*

$e^{t/T} = 1, 1.65, 2.72, 4.48$

(shown by circled points)
$e^{t/T}$ lies close to line segment curve.

2. (a) Plot a curve of $y = 10^x$ by evaluating y at $x = 0$, 1/4, 1/2, 3/4, and 1, and connecting the points with a smooth curve. (If your calculator does not have a 10^x key, use the $\sqrt{}$ key to calculate the fourth and other roots.)

$10^0 = 1$; $10^{1/4} = \sqrt{\sqrt{10}} = 1.78$; $10^{1/2} = 3.2$
$10^{3/4} = (1.78)^3 = 5.6$; $10^1 = 10$.

(b) Estimate the rates of increase of y at several points ($x = 0.2$, 0.4, 0.9) by drawing lines tangent to the curve at those points and calculating the slopes of the lines. Plot these rates of increase on the same graph.

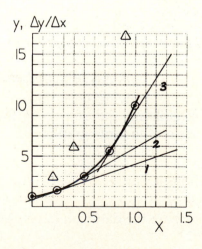

slope1 = 2.0/0.30 = 2.8 ⎫ *Plotted as*
slope 2 = 2.0/0.35 = 5.7 ⎬ *triangular*
slope3 = 5.0/0.30 = 16.7 ⎭ *points.*

(c) Compare the data you have plotted. What is the proportionality constant relating 10^x and its rate of growth? *Ratios of slopes to 10^x: 2.8/1.3 = 2.2 ; 5.7/2.5 = 2.3 ; 16.7/7.5 = 2.2 → average ≈ 2.2 (Theory gives 2.3)*

3. This is essentially a repeat of question 2, except that it focuses on the function $y = e^x$.

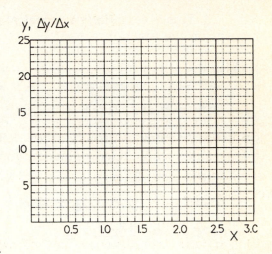

(a) Use the e^x key on your calculator to plot the curve out to x = 3.

(b) Estimate rates of growth at x = 0.5, 1.5, and 2.5 and mark these values with circled points on your graph.

(c) From the plotted information determine how the rates of growth of e^x relate to e^x.

4. An example problem in Review 16 discussed exponential growth of a population given by $p = p_o e^{t/T}$, with $p_o = 1000$ animals and $T = 0.50$ year. The population growth may be also expressed using a power of 2 raised to (t/T_2), where T_2 is the doubling time.

(a) Use the formula $2^x = e^{0.69x}$ to determine the doubling time. Compare with the value of 0.35 year estimated from the curve shown with the example.

(b) What is the population at $t = T_2$? at $t = 2T_2$?

(c) The example gives the population at $t = T$ as 2718 animals. What is it exactly one doubling time later, i.e., at $t = T + T_2$?

5. The intensity of radiation R given off by a radioactive substance decreases according to

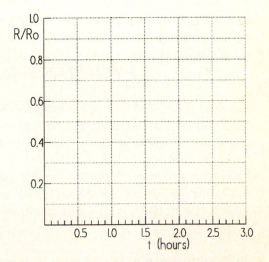

$$R = R_o \, e^{-t/T} , \text{ where } T = 2.0 \text{ hours.}$$

(a) Plot R/R_o out to t = 3.0 hours.

(b) From the curve find the time it takes for the intensity of the radiation to decay to one-half its initial value. (This time is called the half-life $T_{1/2}$.) Compare this with T.

(c) Without using a calculator determine the value of R/R_o at t = $2T_{1/2}$.

3. This is essentially a repeat of question 2, except that it focuses on the function $y = e^x$.

(a) Use the e^x key on your calculator to plot the curve out to x = 3. $e^0 = 1$; $e^{0.5} = 1.7$; $e^1 = 2.7$; $e^{1.5} = 4.5$ $e^2 = 7.4$; $e^{2.5} = 12.2$; $e^3 = 20.1$

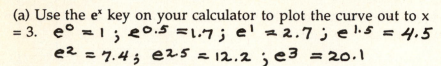

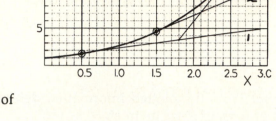

(b) Estimate rates of growth at x = 0.5, 1.5, and 2.5 and mark these values with circled points on your graph.

$slope_1 = 3.0/2.2 = 1.4$
$slope_2 = 7.0/1.6 = 4.4$ } Marked with circled points
$slope_3 = \dfrac{12.0}{1.0} = 12.0$

(c) From the plotted information determine how the rates of growth of e^x relate to e^x. Slope/e^x ratios:

$Ratio 1 = 1.4/1.7 = 0.8$; $Ratio 2 = 4.4/4.5 = 1.0$; $Ratio 3 = 12.2/12.0 = 1.0$

4. An example problem in Review 16 discussed exponential growth of a population given by $p = p_0 e^{t/T}$, with $p_0 = 1000$ animals and T = 0.50 year. The population growth may be also expressed using a power of 2 raised to (t/T_2), where T_2 is the doubling time.

(a) Use the formula $2^x = e^{0.69x}$ to determine the doubling time. Compare with the value of 0.35 years estimated from the curve shown with the example. write expression for p using both bases: $P_0 2^{t/T_2} = P_0 e^{0.69t/T_2} = p = P_0 e^{t/T}$
Thus: $0.69/T_2 = 1/T \longrightarrow T_2 = 0.69 T = (0.69)(0.50 \, yr) = 0.35 yr.$

(b) What is the population at $t = T_2$? at $t = 2T_2$?

at $t = T_2$: $p = 2 P_0 = 2000$ animals; at $t = 2T_2 = 2(2p_0) = 4000$ anima

(c) The example gives the population at t = T as 2718 animals. What is it exactly one doubling time later, i.e., at $t = T + T_2$? In a doubling time, population increases by factor of 2: $p = 2 (2718 animals) = 5436$ animals

5. The intensity of radiation R given off by a radioactive substance decreases according to

$$R = R_0 e^{-t/T}, \text{ where } T = 2.0 \text{ hours.}$$

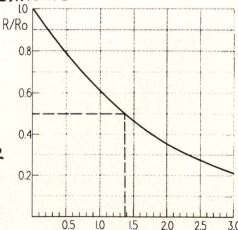

(a) Plot R/R_0 out to t = 3.0 hours. Compute $e^{-t/T}$ at $t = 0, 1, 2, 3$ hr and connect points $e^0 = 1$; $e^{-1/2} = 0.61$; $e^{-1} = 0.37$; $e^{-1.5} = 0.22$

(b) From the curve find the time it takes for the intensity of the radiation to decay to one-half its initial value. (This time is called the half-life $T_{1/2}$.) Compare this with T.

(c) Without using a calculator determine the value of R/R_0 at t = $2T_{1/2}$. It is reduced by half again $\rightarrow$ $R/R_0 = 0.25$

Review 17 — Logarithmic Functions: Big Things, Little Things

Logarithmic functions are the inverse of exponential functions. For example, $\log(10^x) = x$; likewise, $10^{\log x} = x$. In effect, the logarithm of a number is simply the *exponent* to which a base can be raised to give that number. Just as scientific notation makes use of exponents to write very large or very small numbers, so some scientific quantities which vary over a large range are conveniently described in terms of the exponents themselves, i.e., in terms of logarithms. This section reviews how logarithms are manipulated mathematically, and how they are used to scale and graph physical information.

LOGARITHMS TO THE BASE 10

Logs and powers of 10. A good sense of the meaning of logarithms comes from considering how numbers are written in scientific notation. For example the numbers

| 0.01 | 1 | 100 |

can be expressed as follows:

| 10^{-2} | 10^0 | 10^2 . |

The *logarithms to the base 10* (denoted by $\log_{10}$, or simply *log*) of these numbers are the exponents themselves. Thus

$$\log_{10} 0.01 = -2 \qquad \log_{10} 1 = 0 \qquad \log_{10} 100 = 2 .$$

This concept of a logarithm as an exponent can be extended to include non-integer powers as well. In general

> • *the logarithm to the base 10 of a number is the exponent in the power of ten which yields that number,* i.e., $\log_{10} 10^x \equiv \log 10^x = x$.

As suggested in the previous review, the use of exponents which are integers or rational fractions are readily understood, whereas general non-integer exponents require more sophisticated explanation. For simplicity, however, non-integer exponents (logs) can be thought of as values along a continuous curve connecting integer (or fractional) powers. This approach is applied in the following example, in which a typical curve of the logarithmic function, including non-integer values, is plotted.

Plot a curve of $\log x$ ($\log_{10} x$) versus x by connecting points corresponding to $x = 10^{1/2}, 10^{1/3}, 1$, and $10^{-1/2}$, with a smooth curve. From the curve estimate $\log 2$ and $\log \frac{1}{2}$, and compare with values obtained from an electronic calculator.

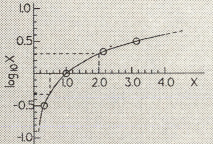

DISCUSSION: Using the basic property $\log 10^x = x$, we obtain:

$$\log 10^{1/2} = \log 3.16 = 1/2$$
$$\log 10^{1/3} = \log 2.15 = 1/3$$
$$\log 10^0 = \log 1 = 0$$
$$\log 10^{-1/2} = \log 0.316 = -1/2 .$$

The corresponding points are plotted above. From the smooth curve connecting the points we estimate $\log 2 = 0.30$ and $\log \frac{1}{2} = -0.30$. A calculator (which uses a sophisticated computational algorithm) yields ± 0.30103 for these logarithms.

Some important particulars. The example just discussed highlights these facts:

Logarithms of numbers in the range 0 to 1 are negative, and log 1 = 0. Also, logarithms are not defined for numbers less than or equal to zero.

These facts apply to all logarithms, no matter what the base (see below). Moreover, in view of the fundamental concept of logarithms as exponents, one should keep in mind that

• *logarithms and the arguments of logarithms are dimensionless numbers.*

Thus, in physics problems, the argument (y in log y) often appears as a ratio or product in which there is a cancellation of units. The next section explains how to deal with logarithms of products and ratios.

The arithmetic of logarithms. (The rules outlined in this section apply equally well to logarithms using other bases.) You will recall from the discussion of scientific notation in Review 1, that powers of ten are multiplied by adding exponents. Thus, in view of the fact that logarithms are actually exponents, the logarithm of a product is

$$log \ (xy) = log \ x + log \ y \ .$$

Likewise, since division of powers is done by subtracting exponents,

$$log \ (x/y) = log \ x - log \ y \ .$$

When the argument of a logarithm is a product of n identical factors, we get this important rule for the logarithms of a power function

$$log \ x^n = n \ log \ x \ .$$

This rule is general; n need be neither an integer nor positive. For instance, the logarithm of a square root $log \ \sqrt{a} = log \ a^{1/2} = (1/2)log \ a$.

No algebraic simplifications exist for the expressions (log a)(log b) or log a/log b. An illustration of how logarithms may be manipulated is the following example:

Sound intensity I can be stated in terms of sound energy flow per unit area, i.e., watts/cm². Since audible intensities vary over such a wide range, however, a logarithmically defined "decibel" unit is used which is based on intensity *relative* to a "threshold" intensity $I_o = 10^{-16}$ W/cm²), as follows:

$$decibels \ (dB) = 10 \ log_{10}(I/I_o) \ .$$

In going from a quiet office to a busy city street, the increase in decibels is about 30 dB. By what factor does the sound intensity (W/cm²) increase?

DISCUSSION: 30 dB = dB(office) - dB(street)
$$= 10 \ log(I_{street}/I_o) - 10 \ log(I_{office}/I_o)$$

(Note that the label "dB" is dropped in the equation which follows. The decibel, like the radian, is a dimensionless unit and is sometimes omitted in a calculation, once the meaning of the numbers is made clear.)

Dividing by 10 on both sides of the equation, and combining logs:

$$3 = \log\frac{I_{street}/I_o}{I_{office}/I_o} = \log\frac{I_{street}}{I_{office}}$$

Taking powers of 10 on both sides yields

$$10^3 = 10^{\log[I(street)/I(office)]} = I_{street}/I_{office} .$$

In other words, the intensity is a factor of 1000 greater in the street.

OTHER BASES

Base 2. Just as $\log_{10}x$ can be thought of as the mathematical inverse of 10^x, $\log_2 x$ is the inverse of 2^x. In other words

$$\log_2 2^x = x \qquad and \qquad 2^{\log_2 x} = x.$$

Base 2 mathematics is important for the study of probabilities, computers, and related subjects. While $\log_2$ is not mentioned often in introductory physics, working some exercises with base 2 is helpful for understanding logs and exponentials. Since $\log_2 x$, like $\log_{10} x$, can be thought of as an exponent, the arithmetic outlined for base 10 holds true for this base (and other bases), as well.

Natural logs. As with the exponential function, there are often mathematical simplifications to be acheived by using e = 2.718.... as the base of the logarithmic function. A number of topics in introductory physics are discussed in terms of the function $\log_e x$ function, called the "natural log" and usually written $ln\ x$. As with other bases, this function is inverse to the corresponding exponential function e^x, so that

$$\ln e^y = y \qquad and \qquad e^{\ln y} = y .$$

Also, as has been already mentioned, the "arithmetic" of natural logs (such as taking the log of a product) is identical to that used with other bases. The "naturalness" of using base e in logarithmic expressions is analogous to the naturalness of using radians when describing oscillations: mathematical derivations are often free of arbitrary constants.

Converting among bases. Usually the conversion of a logarithmic expression to a corresponding expression using another base, if needed, will be worked out for you in an introductory physics text. However, doing so yourself is excellent practice in learning to manipulate logarithmic expressions. Here is one example:

Write an equation which relates $\log x$ to $\ln x$, i.e., $\log_{10}x$ to $\log_e x$.

DISCUSSION: If $y = \ln x$, then $x = e^y$. Thus

$$\log x = \log e^y = y \log e = (\ln x)(\log e).$$

Using an electronic calculator, we find $\log e = \log 2.718 = 0.434$, so that

$$\log x = 0.434 \ln x .$$

Skill Drill 17 — SOLUTIONS AND ANSWERS

1. Without using a calculator, find $y = \log_2 x$ for $x = 2^{3/2}$, 2, $2^{1/2}$, 1, and $2^{-1/2}$.   Plot y versus x on the graph at the right, and connect the points with a smooth curve.

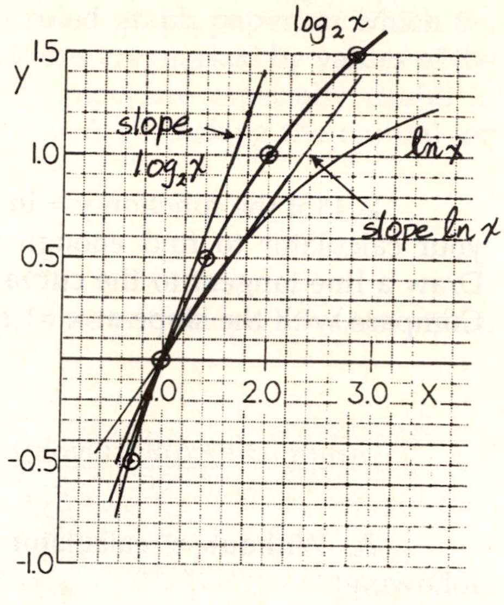

$y = \log_2 2^{3/2} = \log_2 (\sqrt{2})^3 = \log_2 2.83 = 3/2$
$\log_2 2 = 1$
$\log_2 2^{1/2} = \log_2 \sqrt{2} = \log_2 1.41 = 1/2$
$\log_2 1 = 0$
$\log_2 2^{-1/2} = \log_2 \frac{1}{\sqrt{2}} = \log 0.71 = -1/2$

2. Plot the function $y = \ln x$ on the graph at the right, using your calculator to find enough points to draw a smooth curve. Draw a line tangent to the curve at $x = 1$ and determine its slope. Compare with the slope at $x = 1$ for the $\log_2$ curve.

$\text{slope of } \ln x = \frac{1.2 - 0}{2.4 - 1.0} = 0.9$ $\left(\text{Theoretically} = \frac{1}{x} = 1.0\right)$

$\text{Slope of } \log_2 X = \frac{1.2 - 0}{1.8 - 1.0} = 1.5$ $\left(\text{Theoretically} = \frac{\log_2 e}{X} = 1.4\right)$

3. Without a calculator find numerical values for the following:

$\log_2 2 = \log_2 2^1$
$\quad = 1$

$\log_2 4 = \log_2 2^2$
$\quad = 2$

$\ln 1 = 0$

$\log_{10} 1 = 0$

$\log_{10} 0.01 = \log 10^{-2}$
$\quad = -2$

$\ln (1/e) = \ln e^{-1}$
$\quad = -1$

4. Sound level in decibels is related to sound intensity I (W/m^2) by

$$dB = 10 \log_{10}(I/I_0),$$

where $I_0 = 10^{-12}$ W/m^2. Find I for these sounds: (a) jet engine at 160 ft (150 dB); (b) average rock band (120 dB); (c) car interior at highway speed (90 dB); (d) busy street (70 dB).

(a) $150 = 10 \log_{10} I/I_0$
$15 = \log_{10} I/10^{-16} W/m^2$
$I = (10^{-12} W/m^2) 10^{15} = 1000 W/m^2$

(b) $12 = \log_{10} I/I_0$
$I = (10^{-12} W/m^2) 10^{12}$
$\quad = 1 W/m^2$

(c) $9 = \log_{10} I/I_0$
$I = (10^{-12} W/m^2) 10^9$
$\quad = 10^{-3} W/m^2$

(d) $7 = \log_{10} I/I_0$
$I = (10^{-12} W/m^2) 10^7$
$\quad = 10^{-5} W/m^2$

5. The exponentials to the natural base e and to base 2 are related by $e^x = 2^{kx}$. What is k? (You may use the **lnx** key on your calculator.)

6. The natural log and log to the base 2 are related by $k \ln x = \log_2 x$. What is k? (You may use the **lnx** key on your calculator.)

7. Suppose a certain country's population growth during the present century grew according to the following table.

Year	Population (millions)
1900	10
1920	15
1940	22
1960	33
1980	49

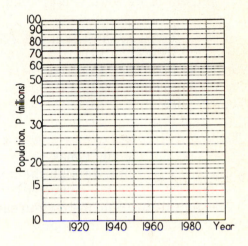

These data can be expressed as an exponential function $P = P_o e^{t/T}$ (a) On the right make a semi-log plot of the data and draw a straight line through the points. (b) From the slope of the line determine the characteristic time T. (c) Assuming the same population trend in previous centuries, approximately how many years ago was the population 1000 persons?

8. The time τ it takes for a pendulum to make a complete swing back and forth depends upon the length L of the pendulum according to a power law: $\tau = A L^n$. Use a log-log plot of the data listed in the table below to find n, as follows:

(a) Plot the data on the graph at the right and draw a straight line through the points. (b) Evaluate the slope of the logarithmic plot and set it equal to n, i.e., $n = (\Delta \log \tau / \Delta \log L)$. Does this agree with the power law given in Review 13? (c) What slope do you get evaluating $(\Delta \ln \tau / \Delta \ln L)$?

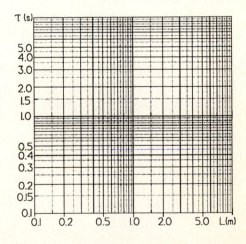

L(m)	τ(s)
0.10	0.6
0.25	1.0
0.55	1.5
1.0	2.0
4.0	4.0

5. The exponentials to the natural base e and to base 2 are related by $e^x = 2^{kx}$. What is k? (You may use the **lnx** key on your calculator.) *Take nat. log of both sides:*

$$\ln e^x = \ln 2^{kx}$$
$$x \ln e = kx \ln 2$$
$$1 = k \ln 2 \longrightarrow k = 1/\ln 2 = 1.44 ---$$

6. The natural log and log to the base 2 are related by $k \ln x = \log_2 x$. What is k? (You may use the **lnx** key on your calculator.) *Let $x = 2$. Then*

$$k \ln 2 = \log_2 2 = 1$$
$$k = 1/\ln 2 = 1.44$$

7. Suppose a certain country's population growth during the present century grew according to the following table.

Year	Population (millions)
1900	10
1920	15
1940	22
1960	33
1980	49

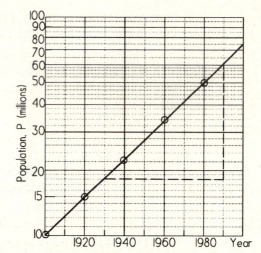

These data can be expressed as an exponential function $P = P_o e^{t/T}$ (a) On the right make a semi-log plot of the data and draw a straight line through the points. (b) From the slope of the line determine the characteristic time T. (c) Assuming the same population trend in previous centuries, approximately how many years ago was the population 1000 persons?

(b) $slope = \frac{1}{T} = \frac{\ln 60 - \ln 18}{(1990 - 1925) yr} = \frac{\ln(60/18)}{65 yr} = \frac{1.2}{65 yr}$

$$T = 54 \, yrs.$$

(c) Taking year 1900 as $t = 0$; thus $P_o = 10 \, million = 10^7$
$$1000 = 10^7 e^{t/T} \rightarrow 10^{-4} = e^{t/T}$$
$$t/T = \ln 10^{-4} = -9.2 \rightarrow t = -(54 yr)(9.2) \rightarrow 500 \, yr \, before \, 1900.$$

8. The time τ it takes for a pendulum to make a complete swing back and forth depends upon the length L of the pendulum according to a power law: $\tau = A L^n$. Use a log-log plot of the data listed in the table below to find n, as follows:

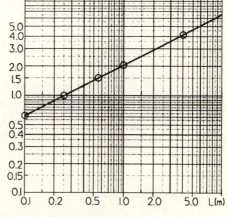

(a) Plot the data on the graph at the right and draw a straight line through the points. (b) Evaluate the slope of the logarithmic plot and set it equal to n, i.e., $n = (\Delta \log \tau / \Delta \log L)$. Does this agree with the power law given in Review 13? (c) What slope do you get evaluating $(\Delta \ln \tau / \Delta \ln L)$?

L(m)	τ(s)
0.10	0.6
0.25	1.0
0.55	1.5
1.0	2.0
4.0	4.0

(b) $n = slope$
$$= \frac{\log 5.0 - \log 1.0}{\log 6.0 - \log 0.25}$$
$$= \frac{\log(5.0/1.0)}{\log(6.0/0.25)} = \frac{0.70}{1.38} = 0.51$$

OK: Review 13 has $T \propto L^{1/2}$

(c) $n = \ln 5.0 / \ln 24 = 1.6/3.2 = 0.50$, *the same.*

Fifth Round Posttest — Optimum test time: 25 minutes or less

This posttest completes the formal work laid out in this book which would help you cope with the demands of a course in general physics in which calculus is not used. If that is the level of your physics course — congratulations! Successful completion of the last five Rounds should give you assurance that you have a mastery of the prerequisite skills needed to deal with problem solving in physics. In addition, you should have some understanding of physics as a discipline: its scope, some important vocabulary, and its approach to discovery. If your physics course requires calculus you will probably want to go on to Round VI after you are satisfied with your work on this test.

Work rapidly but accurately. Check your answers and, if necessary, return to the reviews, drills, and the pretest for further help.

STARTING TIME_____   ANSWERS

1. On the graph at the right sketch in the following curves:

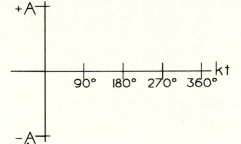

    (a) $y = -A \sin kt$

    (b) $y = A \sin(kt - 45°)$

2. The following identity holds true for a certain phase angle θ_o between 0 and 180°:
$$\cos(\theta + \theta_o) = -\sin \theta .$$

Compare a curve of $\sin \theta$ (as in question 1(a) above) with an appropriately phase-shifted cosine curve, to determine
    (a) θ_o in degrees,   _____

    (b) θ_o in radians.   _____

3. Mount Hood, 11,245 ft high, is located 45 miles ("as the crow flies") from the city of Portland, Oregon. When viewed from sea level in Portland, what is the angle of elevation of the summit in degrees? (Use a small angle approximation.)   _____

4. Substitute $x = e^y$ into the expression $\log_{10} x = 0.434 \ln x$ in order to find a relationship between e^y and a power of 10.   _____

5. Without a calculator find numerical values for the following:
    $\log_2 4$ _____

    $\ln(1/e)^2$ _____

    $\log_{10} 1000$ _____

6. Species of mammals vary in size enormously. Order of magnitude values for mass M and surface area S of some mammals are shown in the table. Fit these data with a straight line on the log-log graph at the right. Determine the exponent in the power law expression $S = AM^n$ by finding the slope of the logarithmic plot $n = \Delta\log S / \Delta\log M$.

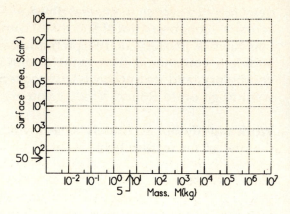

	M(kg)	S(cm²)
Mouse	10^{-2}	10^2
Rat	10^{-1}	3×10^2
Rabbit	10^0	10^3
Human	10^2	3×10^4
Cattle	10^3	10^5
Elephant	10^4	3×10^5
Whale	10^6	10^7

n = _____

ENDING TIME_____

ANSWERS:

1.

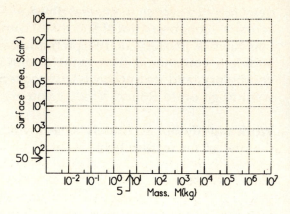

2. (a) 90°
   (b) $\pi/2$ radians

3. 2.7°

4. $e^y = 10^{0.434\,y}$

5. 2, -2, 3

6.

n = 0.6

Essay: What in the World? Particles and Waves

Physics starts out with the most obvious and believable things and moves in patient and logical steps to reveal little-dreamed aspects of the world about us. One of the most obvious ideas about the world is that it is made up of "lumps" of palpable matter which can be separated and combined and moved about in various ways. Physicists sometimes refer to a chunk of matter which occupies a definite region in space and has a certain mass as a "particle." The very first part of almost every physics course is concerned with particle mechanics, the study of how and why rigid chunks of matter move about and rotate in space.

After a while, the concern of a beginning physics course shifts to the consideration of vibratory motion. This part of physics focuses on another obvious aspect of the world — that some motions are repetitive. This is fundamental to an understanding of how such oscillatory motion is propagated from place to place, i.e., the fundamentals of wave motion. (In a wave, shapes or patterns, rather than particles themselves, move from place to place.) Oscillatory and wave phenomena are ubiquitous features of everyday experience — from the "twanging" of a guitar string to the undulations in the surface of a pond of water. To treat these ideas we must expand our study of matter to include not only rigid particles, but deformable objects and materials, such as elastic springs and containers of fluid.

The world as a collection of particles and the world as a symphony of vibrations and waves represent two distinguished themes of scientific inquiry. These concepts are the foundations for describing the connectedness in the world — the ways in which action at one point in space affects conditions in some other place. This dualism characterizes even mundane problems of communication. For instance, suppose you stand outside a friend's house and want to make your presence known. You could toss a pebble (particle) at the friend's window, or you could make a noise (waves); if your friend is overseas you could send a letter (particle) or make a telephone call (waves). Likewise we study the moon either by visiting it or viewing it with a telescope.

Our dual use of the particle and wave description of reality in modern physics echoes some deep scientific concerns. Every one of us at some time probably has confronted a certain philosophical issue about the ultimate nature of reality — perhaps as early as a young child smashing our breakfast cereal into small bits. Each cornflake can be broken into two flakes, and each of these flakes can be broken in two again. Can such subdividing be carried out without limit, giving rise to ever smaller subparticles? Regarding this ultimate question about particles the Greek philosopher Democritus in the 5th century BC pronounced the view, which has since held sway in one form or another, that there is indeed a limit to the possibility of further subdividing. His name for the fundamental unit of matter, the atom (Greek: undivided), is central to scientific vocabulary today, even though it no longer describes the ultimate particle.

For Democritus (and in modified form, for us) the nature of all things is ascribed to the way atoms are arranged. The particle (or atomic) point of view has always had a certain philosophical appeal and has been applied in the explanation of all manner of phenomena. On the other hand, the propagation of sound, while at first confused with the swift ejection of particles of air, was correctly identified in Roman times as spreading undulations in the air set up by vibrating objects like strings and reeds, much like the spreading of ripples on the surface of water. However, for the propagation of light, the choice between particle or wave explanations remained murky, even until relatively modern times. A great deal of optics can be explained in terms of the mechanics of a stream of "corpuscles" (light particles) emanating from luminous bodies. Even Newton cautiously held to that view, depute the enormous speed with which light was recognized to move. After all, it was not clear that light bends around corners (as does sound). Moreover it is difficult to account for the transmission of light from heavenly bodies through apparently empty space. Nevertheless even Newton recognized the evidence for alternative explanations such as the formation of light and dark bands of light reflected from thin films of transparent material, such as might arise from alternating cancellation and reinforcement of waves. But he supposed the bands might somehow be related to vibrations set up by the light corpuscles in the films.

By the late nineteenth century a great clarification had settled in: whereas matter was ultimately a collection of indivisible particles of several types, light was a wave disturbance related to whatever was responsible for the transmission of electric and magnetic forces. Yet within a few years the issue was clouded again. For instance, if we observe with a microscope the surface of a photographic plate or phosphorescent screen lit by weak sources of light, what appears is not uniform, but rather randomly occurring dots of illumination — as if made by *particles* of light. On the other hand, if we carefully watch how electrons (particles) move through or around small openings or obstacles, we don't see clearly defined shadows, but bands of intensity such as would be formed from alternate cancellation and reinforcement of *waves*.

Today physics recognizes that the world has a dual character: particle *and* wave. Which of these predominates depends on the conditions and devices we use to make an observation. And this dual reality is reflected in the choice of fundamental topics with which you begin your study of physics: first particles, eventually oscillations and waves.